A Primer on Theoretical Soil Mechanics

A Primer on Theoretical Soil Mechanics is about adapting continuum mechanics to granular materials. The field of continuum mechanics offers many fruitful concepts and methods, however there is declining interest in the field due to its complex and fragmented nature. This book's purpose is therefore to facilitate the understanding of the theoretical principles of soil mechanics, as well as introducing the new theory of barodesy. This title argues for barodesy as a simple alternative to the plasticity theory used currently and provides a systematic insight into this new constitutive model for granular materials. This book therefore introduces a complex field from a fresh and innovative perspective using simple concepts, succinct equations and explanatory sketches. Intended for advanced undergraduates, graduates and PhD students, this title is also apt for researchers seeking advanced training on fundamental topics.

Dimitrios Kolymbas is Emeritus Professor at the University of Innsbruck, Austria, where he worked from 1994 to 2017 as Professor at the Unit of Geotechnical and Tunnel Engineering. He is the author of several books and many papers, and invented the new constitutive theories of hypoplasticity and barodesy.

A Primer on Theoretical Soil Mechanics

DIMITRIOS KOLYMBAS
University of Innsbruck

Shaftesbury Road, Cambridge CB2 8EA, United Kingdom

One Liberty Plaza, 20th Floor, New York, NY 10006, USA

477 Williamstown Road, Port Melbourne, VIC 3207, Australia

314–321, 3rd Floor, Plot 3, Splendor Forum, Jasola District Centre, New Delhi – 110025, India

103 Penang Road, #05–06/07, Visioncrest Commercial, Singapore 238467

Cambridge University Press is part of Cambridge University Press & Assessment, a department of the University of Cambridge.

We share the University's mission to contribute to society through the pursuit of education, learning and research at the highest international levels of excellence.

www.cambridge.org
Information on this title: www.cambridge.org/9781009210331

DOI: 10.1017/9781009210348

First published 2022

Printed in the United Kingdom by TJ Books Limited, Padstow Cornwall

A catalogue record for this publication is available from the British Library.

ISBN 978-1-009-21033-1 Hardback

Additional resources for this publication at www.cambridge.org/primer.

Contents

Preface

> Socrates taught us to distinguish between what we understand and what we do not understand.

Theoretical soil mechanics is continuum mechanics adapted to soil, a granular material. Continuum mechanics is a complicated, subtle and difficult science. It suffers from fragmentation; its notation is not unique, and the degree of mathematical abstraction varies greatly. Some researchers get lost in it. The many diverse numerical applications add to the confusion; their continuum mechanical foundations are often hidden.

These aspects imply a declining interest in continuum mechanics for soils, which is also enhanced by the emergence of computer-based micromechanical methods. Instead of the continuum, agglomerates of individual particles are considered. This approach has its merits but still does not provide a global understanding for the behaviour of soil. Certainly, the continuum is an abstraction, just like the mass point of classical mechanics. The discontinuous structure of matter (such as sand) can be reconciled with the assumption of the continuum by considering the continuously distributed field quantities as expected values of discontinuous distributions.

The main purpose of this book is to awaken the reader's interest towards the elegant field of continuum mechanics and its applications in soil mechanics. An important part therein is the constitutive law of soils, now in a completely new frame – the theory of barodesy.

Every science starts from its fundamentals and proceeds to ever higher peaks of increasing complexity and refinement. Thus, the author of a textbook that is intended to be clear and concise faces the questions of 'where to start?' and 'where to stop?'. Clearly, the answers can only be subjective in the hope that the reader can retrieve the missing information in both directions. A good book should rather motivate the reader towards a search in literature and in one's own research.

Deeply convinced that 'less is more', I should like to present the reader with a guide to understanding, rather than a repository of equations. I have also attempted to find new and simpler access to several items. In doing so I have followed Walter Noll's suggestions for the role of the professor:

> 'The professor's focus is on understanding, gaining insight into, judging the significance of, and organizing old knowledge. He is disturbed by the pile-up of undigested and ill-understood new results. He is not happy until he has been able to fit these results into a larger context. He is happy if he can find a new conceptual framework with which to unify and simplify the results that have been found by the researcher.'

1 Granular Materials as Soft Solids

1.1 Soil and Geotechnical Engineering

Our solid underground consists of soil and rock; soil being the more important, as our cities are mainly built on it. One usually considers the underground as fixed, and thus, confidently introduces the load of buildings into the soil. However, engineers gradually realised that soil is a rather soft solid that can be easily deformed (Fig. 1.1). Deformation of soil matters as it can lead to settlement and cracks in buildings. Even worse, inclined soil in slopes can move downhill. This motion can be either slow or fast. In the latter case we have catastrophic landslides that can cause thousands of casualties (Fig. 1.2). Landslides can also occur underwater. Gently inclined submarine slopes comprising thousands of cubic kilometres can suddenly start moving giving rise to mega-tsunamis (e.g. the prehistoric Storegga landslide in the North Atlantic). Thus, soil can behave as a fluid, despite its ability to permanently sustain shear stress. Also, horizontally layered soil, i.e. not inclined soil, can be suddenly transformed into a fluid. This is the case when a water-saturated loose sand deposit undergoes a sudden mechanical excitation (e.g. earthquake). The results are peculiar, buildings can sink into the liquefied sand. Sand can also fly (Fig. 1.3). Jet winds can carry thousands of tons of fine sand to heights of up to several kilometres and move it from, say, the Sahara to Europe. Wind is also responsible for the motion of sand dunes.

Exploiting the softness of soil, geotechnical engineers may intervene applying many operations to it. They undertake big excavations to build, e.g. underground garages, or to extract ore or lignite from the underground. They raise earth dams, which can be destroyed by internal erosion if not properly densified. They improve the bearing capacity of foundations by densification of the underground or by the installation of piles. They support cuts or fills in the underground by retaining walls, etc. For all this, one needs to understand the mechanical behaviour of soil and to this end a large variety of experiments have been carried out over the last decades. Being considerably softer than the usual solids (such as steel or concrete), soil samples allow large and complex deformation in the laboratory. They demonstrate a mechanical behaviour that appears extremely complex on first look. Knowledge of this behaviour however, opens the possibility to assess the deformation (i.e. behaviour under loading) and the stability of soil, and thus, preventing catastrophes such as landslides.

Figure 1.1 Footstep on the soil of the moon (NASA). The irreversible deformation of soil manifests its inelastic nature and also its memory. Soil, in particular sand, was the first material with memory exploited by mankind.

1.2 Granulates in Chemical Engineering

Soil is by far not the only granular material of technical relevance. Chemical engineering considers a vast amount of other granular materials such as flour, sugar, coffee beans and ground coffee, soya beans, cement, ore, pellets, etc., which are of high economic importance (Fig. 1.4). Their mechanical behaviour is dictated by their granular nature which is exactly the same as that of soil. Of technical importance is their storage and transportation, the first accomplished in silos. Here, their granular nature poses some difficulties, especially at the outlets of silos, which often get clogged. As for the transportation of granulates, various techniques have been invented, among them dense and dilute phase pneumatic conveying. These bear similarities with the motion of sand in moving dunes and by jet winds.

1.3 Can We Consider Granular Media as Continua?

Should one treat soil as a continuum or rather as a 'discrete' medium composed of individual particles? In the era of digitalisation, there is a tendency to discretise everything. Also, in soil mechanics too, an increasing number of researchers turn to the discrete approach, as the increasing power of computers allow one to consider grains in large numbers. The idea that grains are the truth, and continuum is merely a fiction is gaining traction. This is a deep ongoing philosophical question: what is truth and what is fiction? In physics, scientists are accustomed to accepting a dual approach to tiny corpuscles, considering them both particles and waves. The problem is thus reduced to which method is more appropriate. The advantages of the continuum approach become clear when we recall the saying that there are those unable to see the forest for the trees. In view of the progressing oblivion of

Figure 1.2 Soil can flow: A mure has covered a vehicle. Courtesy Mag. G. Obwaller, Community Wald im Pinzgau.

the continuum mechanics approach, this book aims to help interested scholars gain insight into it. The biggest merit of continuum mechanics is to allow the application of the powerful tool of calculus.

1.4 Differences between Granulates and Other Solids

Contrary to metals and other solids, granulates have a nearly vanishing tensile strength. They only have shear strength, which is mainly of frictional nature, i.e. proportional to normal stress (see Chapter 9). The part of shear stress that is independent of normal stress is called cohesion. A lengthy dispute ensues in soil mechanics as to whether cohesion should be attributed to electromagnetic attraction between the individual grains (so-called true cohesion) or not [92]. There is no conclusive answer as yet, but the author concurs with Schofield [92] that there is (almost) no true cohesion, and that an apparent cohesion is mainly due to interlocking between grains (caution, 'interlocking' here means merely that the grains are toothed and not that they are interwoven). This interlocking causes dilatancy, giving rise to suction in water-saturated soil, such that in the end the strength by cohesion is also frictional.

The other important peculiarity of granulates is the large range of density variation. One and the same soil can be encountered, at the same pressure, in dense and loose state, the latter being a bad underground for foundations. As such, there is no unique relation between pressure and density. In other words, the same density can prevail at different pressures. The pronounced variability of density gives rise

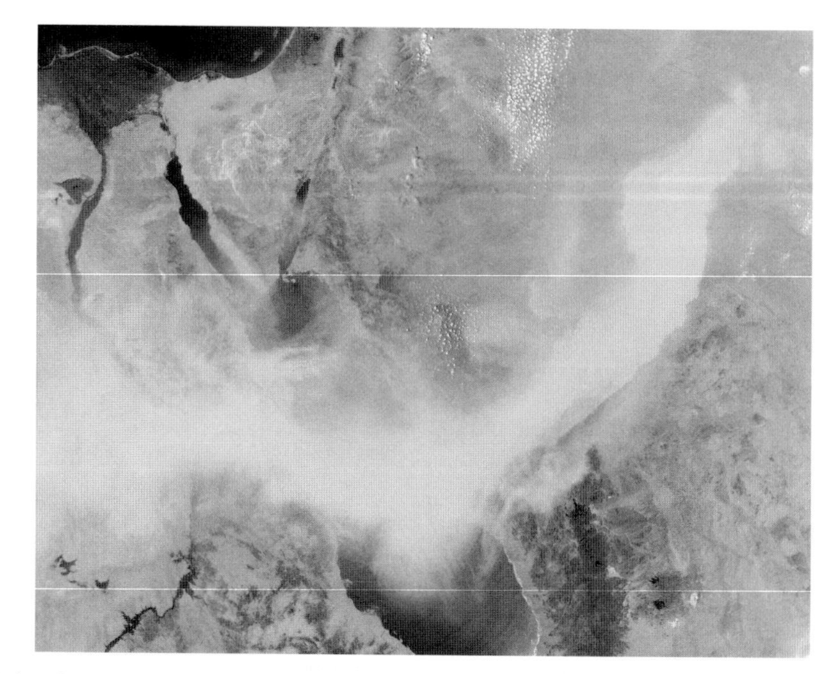

Figure 1.3 Soil can fly: Sand clouds over the Red Sea (NASA).

Figure 1.4 There are many different granular materials, such as (a) spices and (b) gravel.

to the phenomena of dilatancy and contractancy with important implications for water-saturated sandy soil: vibrations can easily transform it into a liquid and this liquefaction is a feared side effect of earthquakes.

2 Mechanical Behaviour of Soil
Experimental Results

2.1 The Meaning of Mechanical Behaviour

The mechanical behaviour of a material is the way it responds to deformations. Herein, the response is expressed as stress. We will therefore consider in this book, the strains and stresses that develop during particular loadings. More specifically, we will consider strain and stress paths in the corresponding strain and stress spaces, as well as the correspondence of stress and strain, the so-called stress–strain curves. Taken that stress and strain are tensors with six independent components each, it appears hopeless to get any insight to the underlying processes. Fortunately, there are special cases wherein the symmetries of these will allow us to consider only two principal stress and strain components, and these cases are sufficient to reveal the main aspects of the mechanical behaviour of soil.

The manifold behaviour of soil (and other granular materials) is investigated in the laboratory by several tests. Soil mechanics comprises many laboratory tests, e.g. grain size analysis by sieving. The *mechanical behaviour* of soil is revealed by stress–strain relations and stress paths obtained with deformation of soil samples.

2.2 Element Tests

Stress and strain cannot be measured directly. One may only measure forces acting upon a soil sample and displacements imposed upon the boundaries of the sample. Therefore, one may only infer the stress (as force divided by area) and strain (as elongation divided by length), provided that these quantities are *constant* in the sample. To this end, we need *homogeneous* deformation, i.e. constant deformation throughout the sample. Sources of inhomogeneity can be initial scatter of density, or shear stresses due to friction along rough walls. Such disturbances can be suppressed by improving experimental techniques. More intricate are inhomogeneities that set on spontaneously. They originate from the simple fact that we deform a sample by imposing displacements to its boundaries, but we cannot enforce the distribution of displacements (and stress) *within* the sample. The problem behind is a mathematical one and refers to the loss of uniqueness of the underlying initial boundary value problem (see Chapter 17). The importance of this question is huge and refers not only to the evaluation of laboratory tests but also to the numerical simulation of geotechnical problems.

The ideal case of a test with homogeneous deformation is called an element test. An element test should not be confused with a real test in the laboratory. Rather, it refers to the deformation of a material point and reveals material properties cleared from any system instabilities, such as the onset of inhomogeneous deformation and shear bands. The outcomes of element tests agree to the ones of real tests as long as the deformation of the latter is homogeneous. It is generally believed that, with some care, laboratory tests initially exhibit homogeneous deformation and preserve it until nearly the peak (see Section 2.5). Recent investigations [19, 108] however indicate that the departure from homogeneity sets on much earlier.

2.3 Typical Laboratory Tests

The most widespread tests to explore the mechanical behaviour of a soil are the oedometric and the triaxial tests. Other tests being the shear-box, the simple shear and the true triaxial tests. In the oedometric and triaxial tests, a cylindrical soil specimen is compressed in a vertical direction. Laterally, either the displacement is inhibited (oedometer, kinematical boundary condition, $\varepsilon_2 \equiv \varepsilon_3 = 0$) or the lateral stress is kept constant (triaxial test, static boundary condition, $\sigma_2 \equiv \sigma_3 = \text{const}$). The two different boundary conditions imply a completely different mechanical behaviour; in the oedometric test the stress increases limitlessly, whereas in the triaxial test a limit stress state is reached.

2.4 Oedometric Test

The inhibition of lateral displacement is achieved by a stiff metallic ring (Fig. 2.1). The soil grains move in a vertical direction along this wall, and this evokes shear stresses that disturb the homogeneous distribution of stress. To keep this effect small, the samples are flat, i.e. their diameter is considerably larger than their height. In the course of the test, the vertical load is increased, and the corresponding settlement of the upper plate is monitored. The results are plots of vertical strain ε_1 (usually plotted downwards) versus vertical stress σ_1 or versus $\log \sigma_1$. Instead of ε_1, the compaction of the sample can be expressed as reduction of the void ratio e or of the so-called specific volume $1 + e$. The void ratio e is defined as the ratio V_v/V_s, with V_v and V_s being the volumes of voids and solids, respectively, within a small but representative

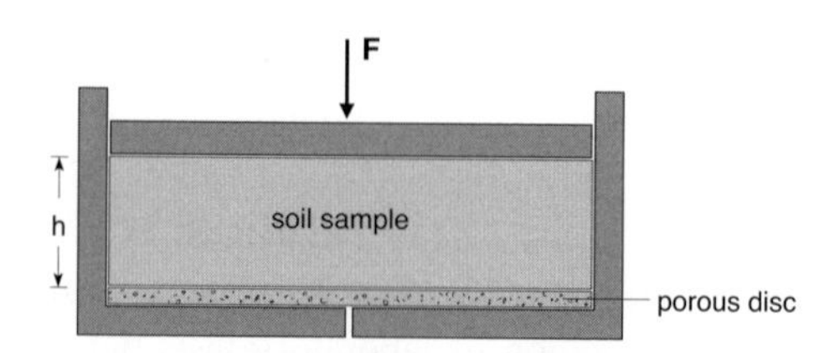

Figure 2.1 Oedometer (schematic). The soil sample is compressed in a vertical direction, whereas lateral expansion is inhibited by a rigid containment. Reproduced from [57], courtesy of Springer Nature.

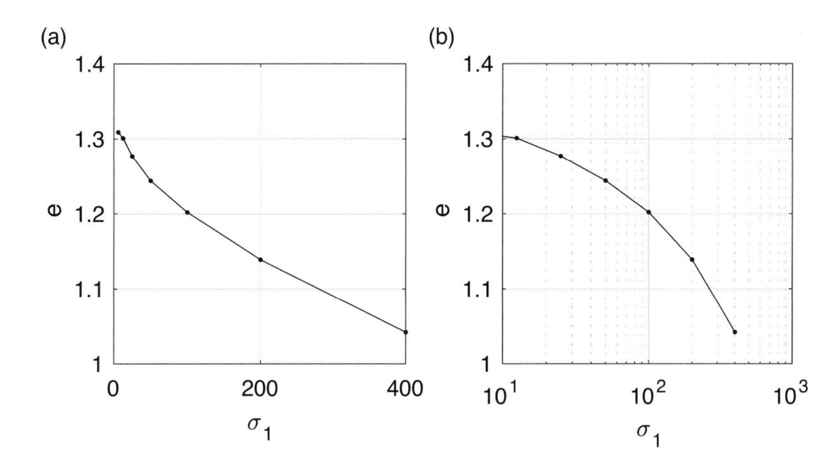

Figure 2.2 Oedometric compression of a clay sample (data from Wichtmann [115]). (a) Linear and (b) semilogarithmic plots. Note the nearly horizontal inclination of the initial part of the curve in the semilogarithmic plot.

volume element. Semilogarithmic plots have to be interpreted with caution, because they change the curvature of the curve: a steep curve becomes nearly horizontal in a semilogarithmic plot and this gives the wrong impression of a high stiffness (Fig. 2.2).

The plots ('compression diagrams') can be approximated equally well either with a logarithmic function

$$e = e_0 - C_c \ln(\sigma_1/\sigma_{10}) \tag{2.1}$$

or with a power function

$$e = (1 + e_0)(\sigma_1/\sigma_{10})^{\alpha} - 1, \tag{2.2}$$

with e_0 and σ_{10} being the values at the beginning of the test. Sudden increases of the settlement can be attributed to grain crushing or breakage of bonds between grains; these bonds making the difference between *natural* and *reconstituted* soil samples. In natural samples, minute bondings between the grains may increase stiffness (Fig. 2.3).

A supposed bend of the compression curve of undisturbed samples due to the transition from reloading to virgin loading is often attributed to the geological preload of a soil deposit, e.g. due to glaciers in the past. However, this 'geologic bend' is often hard to identify and seems to be rather a misconception emanating from plasticity theory (transition from elastic reloading to plastic virgin loading).

The lateral stress σ_2 can only be monitored if the elongation (assumed as small) of the lateral wall is measured ('soft-oedometer' [45]). At loading, σ_2 increases proportional to the vertical stress, $\sigma_2 = K_0\sigma_1$, but at subsequent unloading it decreases much less than σ_1 (Fig. 16.11).

2.5 Drained Triaxial Test

As in the oedometric test, the sample is compressed in a vertical direction. Laterally it is supported by a hydrostatic pressure that is usually kept constant during the

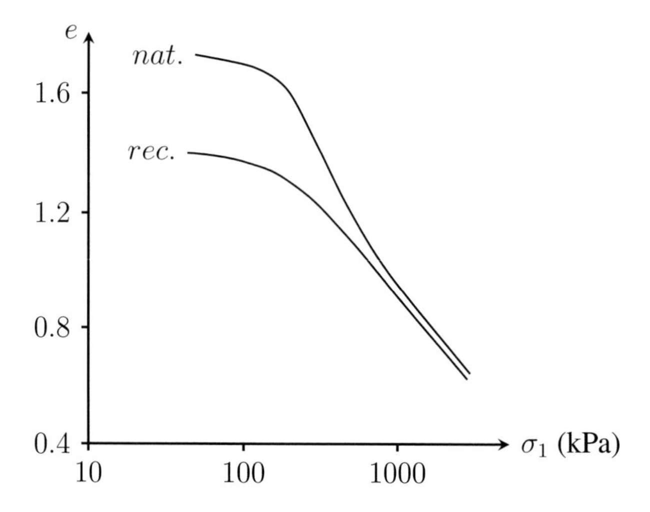

Figure 2.3 Difference between *natural* and *reconstituted* samples, reproduced from [11]. The cementation of an undisturbed (or natural) sample breaks and the compression curve approaches gradually the one of a reconstituted sample.

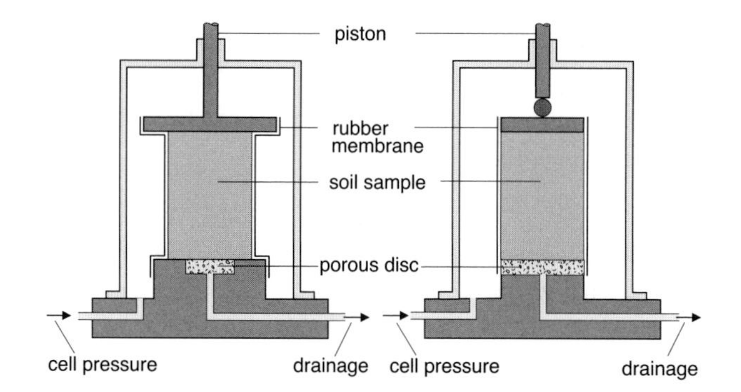

Figure 2.4 Two versions of triaxial tests (schematic). Reproduced from [57], courtesy of Springer Nature.

test: $\sigma_2 = \sigma_3 =$ const (Fig. 2.4). Contrary to the oedometric test, this is not a kinematical but a statical boundary condition and this implies that the sample can laterally expand. Depending on the amount of this expansion, the volume of the sample is increased or decreased.

The triaxial test reveals rich information on the behaviour of a soil. The resulting plots show

The stress path. The following variables are used:

$$\begin{array}{llll} y\text{-axis:} & \sigma_1, & q := \sigma_1 - \sigma_2, & t := \frac{\sigma_1 - \sigma_2}{2}, \\ x\text{-axis:} & \sigma_2, & p := \frac{\sigma_1 + 2\sigma_2}{3}, & s := \frac{\sigma_1 + \sigma_2}{2}. \end{array}$$

The conventional triaxial test ($\sigma_2 =$ const) is represented as a straight line with inclinations 45° in the s–t-plot, and $\arctan 3 \approx 71.6°$ in the p–q-plot.

The stress–strain curve. The vertical strain is given as $\varepsilon_1 := \Delta u_1/h_0$ or as logarithmic strain: $\epsilon_1 := \ln(h/h_0) = \ln(1 + \varepsilon_1)$. h_0 and h are the initial and actual heights of the sample, respectively. Usually, compression is taken as positive. The stress is represented by several variables: σ_1, $\sigma_1 - \sigma_2$, σ_1/σ_2, $\eta := q/p$, $\frac{\sigma_1 - \sigma_2}{\sigma_1 + \sigma_2} = \sin \varphi_m$,

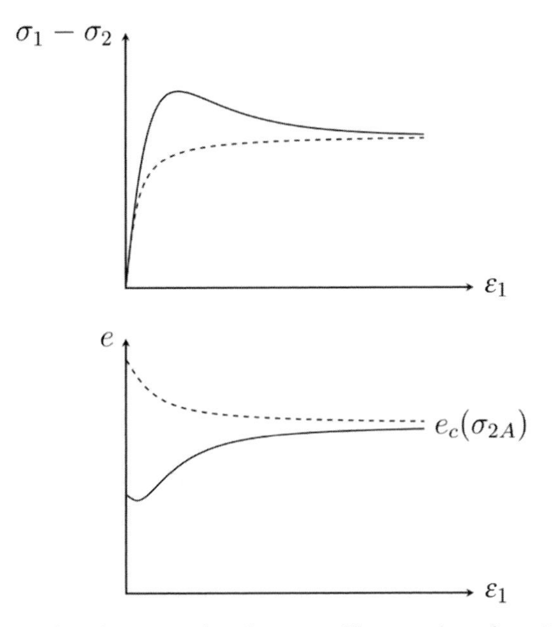

Figure 2.5 Triaxial tests with sand conducted at constant lateral stress σ_{2A}. The curves show schematically the influence of the initial void ratio e ('pyknotropy'). In the course of the tests, the void ratio approaches asymptotically its critical value e_c, which depends on the lateral stress σ_{2A}.

where φ_m is the mobilised friction angle. Depending on the density of the sample, the stress–strain curves either rise to a peak and then decrease ('softening') to a residual value or strive directly to the residual value.

The volumetric strain curve. With full saturation of the pore space, and assuming incompressibility of water and grains, the change of volume is measured by the volume of water squeezed out or sucked into the sample. The volumetric strain $\varepsilon_v = \Delta V/V$ is plotted over the axial strain ε_1. Alternatively, the void ratio e is plotted over ε_1 (Fig. 2.5). Volume increase is taken as positive. Depending on whether the initial density is low or high, e decreases (contractancy) or increases (dilatancy). Inevitably, these volume changes must be limited and eventually lead to a stationary value, the so-called critical void ratio e_c. Higher lateral stresses σ_2 suppress dilatancy and also the peak ('barotropy').

The stress–strain curves and volumetric strain curves are in principle similar for sand (Figs. 16.6–16.8) and clay (Figs. 2.6 and 2.7).

Triaxial extension tests are also conceivable. There, the stresses σ_1 and σ_2 are still compressive, but the sample becomes slenderer, the ratio σ_1/σ_2 is reduced below 1, until a limit state of vanishing stiffness is reached. In compression we have $\sigma_1 > \sigma_2$, whereas in extension it is $\sigma_1 < \sigma_2$.

Sooner or later in the course of a triaxial compression test, shear bands and/or bulging set on as modes of inhomogeneous deformation (Fig. 2.8). The old

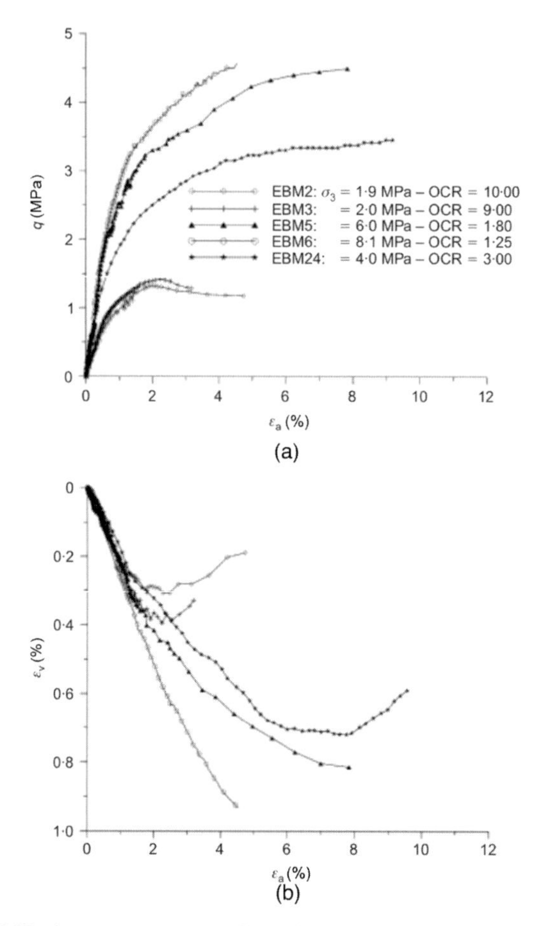

Figure 2.6 (a) Stress–strain and (b) volumetric strain curves of triaxial samples of Boom clay, isotropically consolidated to 9 MPa. Reproduced from [100].

conception that a shear band is formed in dense sand samples, while in loose samples the deformation remains continuous, was refuted by investigations of Desrues et al. [17], who discovered with X-ray tomography that even within loose samples a rather confusing system of shear bands is formed, which externally gives the impression of continuous deformation. In extension tests, the prevailing inhomogeneity is necking (Fig. 2.9).

2.5.1 Barotropy and Pyknotropy

Triaxial tests can be carried out with different lateral stresses $\sigma_2 \equiv \sigma_3 = \text{const}$ and with various initial densities. So-called deviatoric tests are also possible, they are carried out keeping the sum $3p = \sigma_1 + \sigma_2 + \sigma_3$ constant.

Increasing the cell pressure σ_2 leads to higher $\sigma_{1,max}$-values and lower dilatancy. However, the peak friction angle (i.e. arcsin $\varphi_p = (\sigma_1 - \sigma_2)/(\sigma_1 + \sigma_2)_{max}$) decreases. This effect is called *barotropy*.

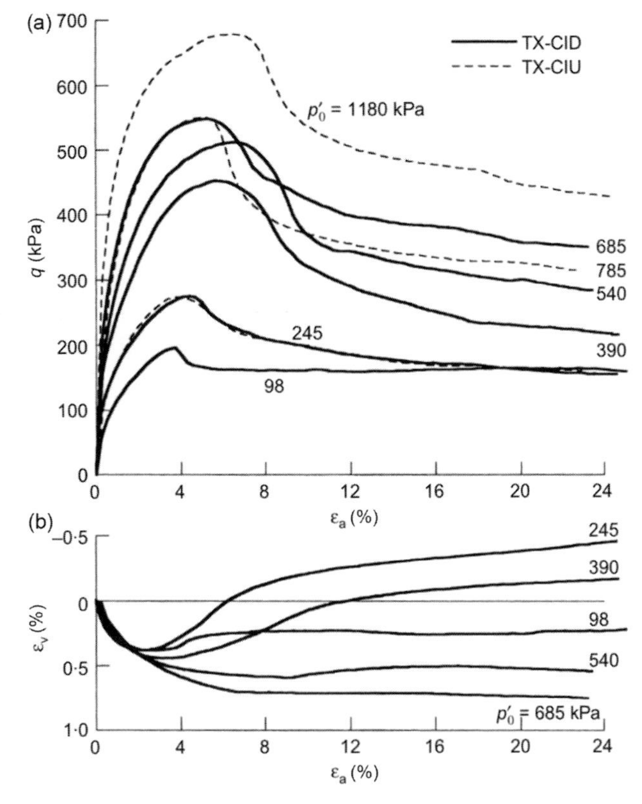

Figure 2.7 (a) Stress–strain curves and (b) Volumetric Strain Curves for standard undrained ('CIU') and drained ('CID') triaxial compression tests on samples of natural Pietrafitta clay, tested at various values of mean effective stress p_0. Reproduced from [12].

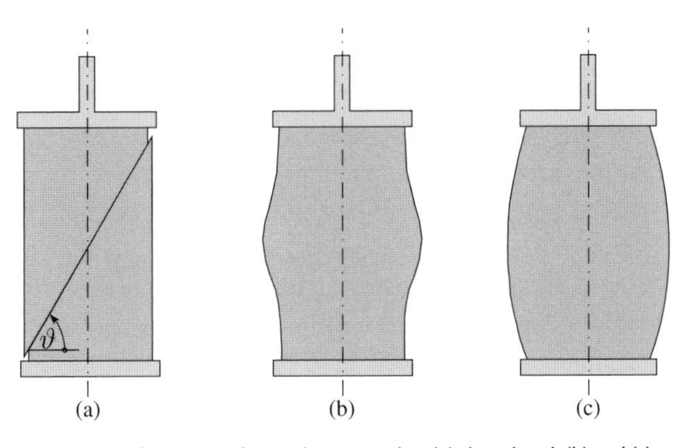

Figure 2.8 Patterns of inhomogeneous deformation of triaxial test samples: (a) shear band, (b) and (c) are two types of barrelling. Reproduced from [57], courtesy of Springer Nature.

In addition, increasing the initial void ratio e of the sample decreases the peak. Upon further increase of $e_{initial}$, the peak vanishes and the stress–strain curve heads directly to the residual value. The influence of e is called *pyknotropy* (Fig. 2.5).

Figure 2.9 Necking of a clay sample, the typical failure pattern of triaxial extension (Courtesy Professor J. Grabe, Technical University Hamburg-Harburg).

2.5.2 Softening

The stress–strain curves of drained triaxial tests can be plotted as graphs of stress deviator $q=\sigma_1-\sigma_2$ versus the axial strain ε_1. Often, they exhibit a *peak*, i.e. a maximum that is followed by a decreasing part of the curve which finally attains a stationary (i.e. constant) value. This behaviour can be easily explained by considering the corresponding volumetric curve ε_v versus ε_1: at peak prevails dilatancy, which however loosens the sample. Inevitably, loosening decreases the maximum sustainable shear stress (or stress deviator). Thus, the peak of a dilatant sample is inevitably followed by softening, and finally a residual state is reached characterised by constant p and constant ε_v. However, softening is also related with a large deformation of the sample, and at this stage inhomogeneities of the deformation are more likely to set on. Hence, softening is often accompanied (and enhanced) by inhomogeneous deformation such as barrelling and/or appearance of shear bands. As a consequence, the softening branch of the stress–strain curve represents a less reliable part of the experiment. Consider, for example, a barrelled sample. To obtain the axial stress σ_1, the measured axial force should be divided by the cross-sectional area of the sample. However, a barrelled sample does not have a unique cross section.

Barrelling, as a pattern of inhomogeneous deformation, was early recognised as disturbing so that efforts have been undertaken to suppress it by using specimens with lubricated end plates and/or squat specimens. However, this has barely helped, see also Chapter 17.

2.6 Undrained Triaxial Tests

A water-saturated sample is first hydrostatically loaded. Drainage is then inhibited, and the sample is compressed in axial direction with constant lateral (total) stress σ_2.

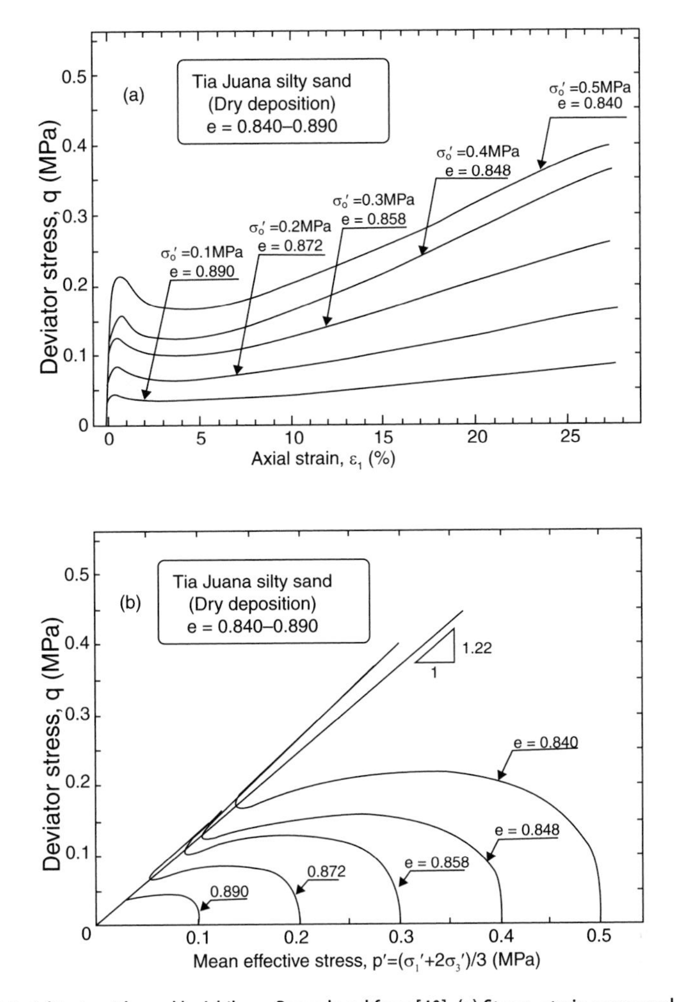

Figure 2.10 Undrained triaxial tests with sand by Ishihara. Reproduced from [40]. (a) Stress–strain curves and (b) stress paths.

As the pore water and the grains are practically incompressible, the condition of constant volume is achieved by the fact that the pore water cannot be squeezed out. Consequently, the pore pressure increases and can be measured by pore pressure transducers. The effective stress path (see Section 19.3) depends on the density (or void ratio) of the sample and has an altogether different shape than in the drained case (Figs. 2.10–2.13). In a q versus p diagram (p refers here to the effective stresses), it is curved and the shape of the curve depends on the void ratio and also on the method of sample preparation (Fig. 2.14). It is initially curved and approaches asymptotically the critical state line $q = mp$. An initial direction towards decreasing p-values indicates the behaviour of a loose (or contractant) soil. At some point however, the stress path turns more or less suddenly towards increasing p-values and the sample now exhibits the behaviour of a dense (or dilatant) soil. This turn is called phase transformation. The stress–strain or q versus ε_1 curves are also peculiar. Their slope is not monotonic. Some of them exhibit a peak, which is then followed by softening and then again by a further increase of the deviator q, see Figs. 2.10 and 2.12.

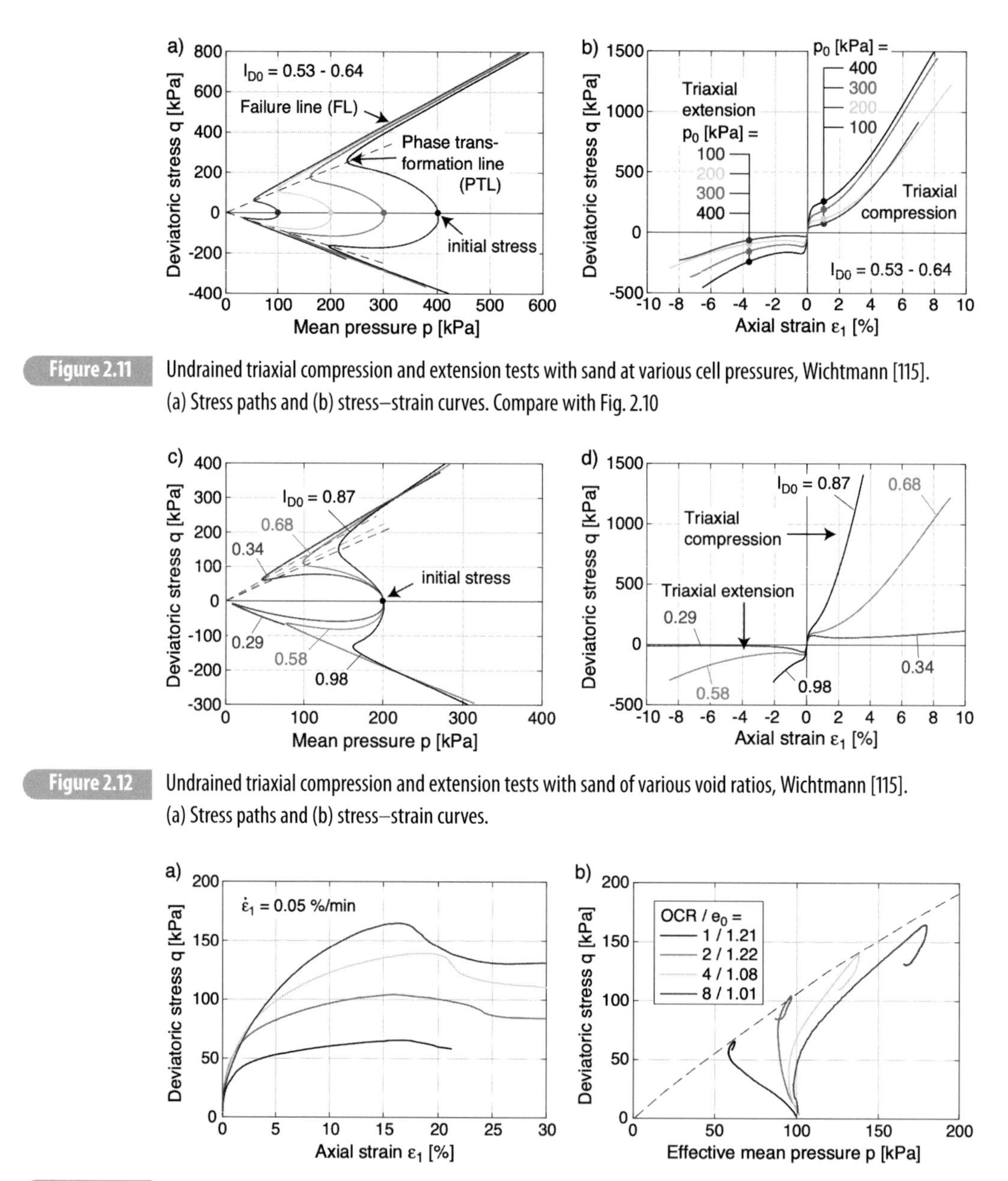

Figure 2.11 Undrained triaxial compression and extension tests with sand at various cell pressures, Wichtmann [115]. (a) Stress paths and (b) stress–strain curves. Compare with Fig. 2.10

Figure 2.12 Undrained triaxial compression and extension tests with sand of various void ratios, Wichtmann [115]. (a) Stress paths and (b) stress–strain curves.

Figure 2.13 Undrained triaxial tests with clay of various void ratios (Wichtmann [115]). (a) stress-strain curves (b) stress paths.

2.7 Cyclic Tests

Cyclic loading appears often, e.g. due to wind and waves in offshore foundations, or due to trains, etc., and intensively concerns soil mechanics, as the corresponding behaviour of soil exhibits a number of special features. According to a now obsolete concept, an unloading-reloading cycle is elastic, i.e. it does not produce any additional deformation. This is however not true. In reality such a cycle produces an increment of deformation and the question referring to consecutive cycles is whether

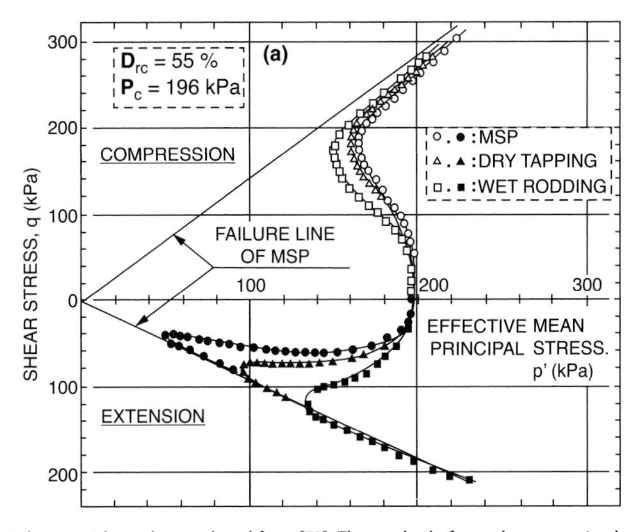

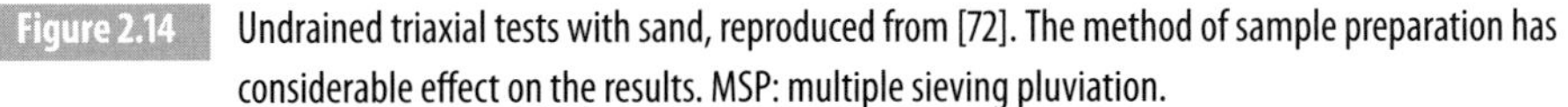

Figure 2.14 Undrained triaxial tests with sand, reproduced from [72]. The method of sample preparation has considerable effect on the results. MSP: multiple sieving pluviation.

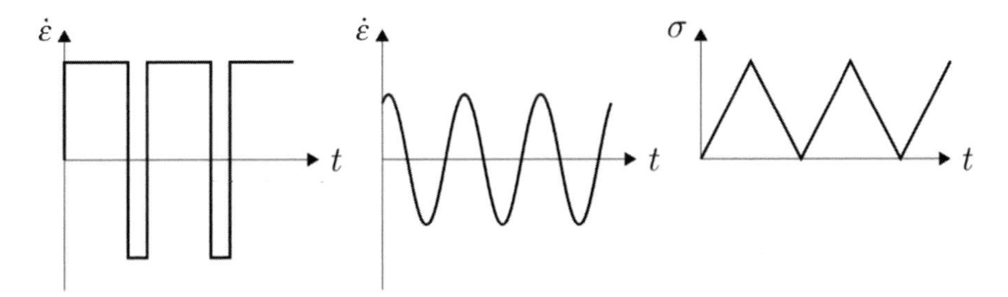

Figure 2.15 Possible loading programmes for cyclic tests.

the series of these increments is bounded (so-called shake-down) or unbounded (so-called incremental collapse). The loading can be strain- or stress-controlled (see Section 2.10), which implies that the switch from loading to unloading and vice versa is defined in terms of stress or strain. Fig. 2.15 shows some possible loading programmes for cyclic tests.

In the case of oedometric loading, the series of incremental deformations (Fig. 2.16) is of course bounded but still adverse. Railway companies, for example, spend huge amounts of money every year to tamp their track beds to counteract the progressive subsidence caused by the repeated passage of trains.

Regarding cyclic triaxial tests, we have to distinguish between drained and undrained ones and also between cyclic stress and cyclic strain. Clearly, the initial values of stress and void ratio play a role as well as the amplitude of the cycles.

The results of a cyclic *drained* triaxial test are shown in Fig. 2.17: at each cycle the stress deviator is reduced to 0. Upon reloading, the stress–strain curve rises more or less exactly to the point of unloading.

Undrained cyclic triaxial tests are important for the study of liquefaction, which is related to substantial damage of buildings in the case of earthquakes. Undrained cyclic *strain* can (depending on void ratio) lead to the zero-stress point where the stiffness vanishes (so-called total liquefaction), see Fig. 2.18. Undrained cyclic *stress* leads to stress paths that oscillate between the prescribed limits along butterfly-shaped patterns. The corresponding strain oscillates between progressively

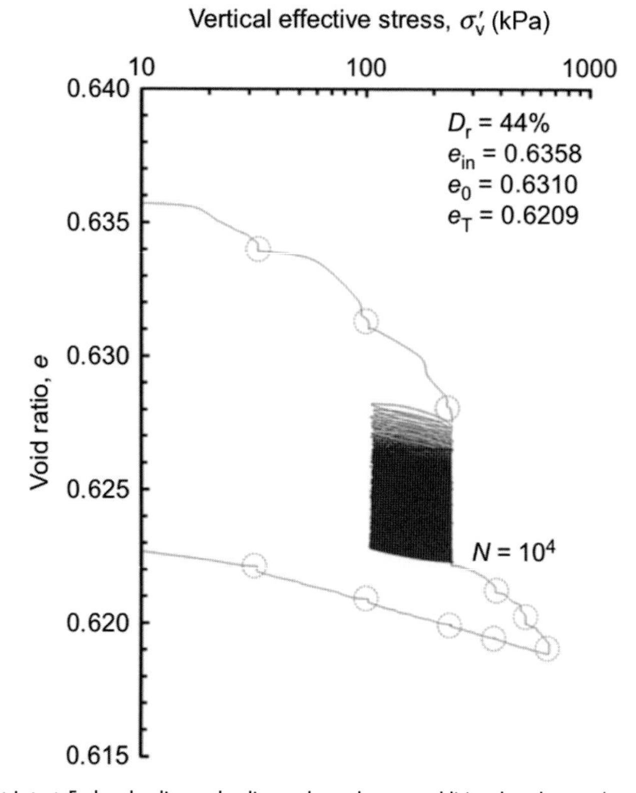

Figure 2.16 Cyclic oedometric test. Each unloading–reloading cycle produces an additional settlement (or reduction of void ratio e). Reproduced from [78].

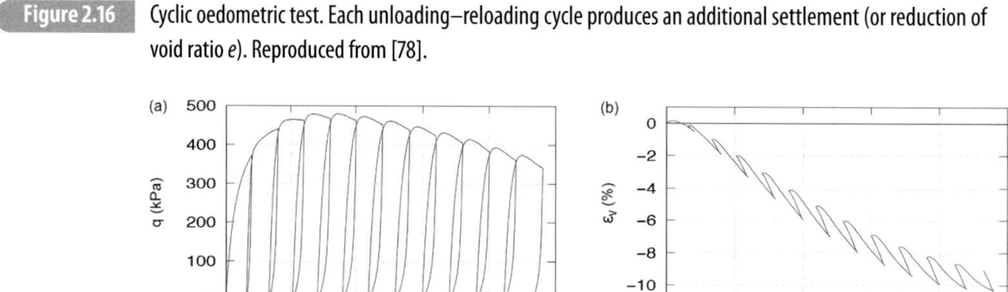

Figure 2.17 Drained triaxial tests with sand by Wichtmann [115]. (a) Stress–strain curve and (b) volumetric strain curve at repeated loading and unloading.

widening limits. This sort of incremental collapse is called 'cyclic mobility', see Figs. 2.19 and 2.20.

2.8 True Triaxial Apparatus

The true triaxial apparatus allows rectilinear extensions with independent control of displacement in all three directions. This appears at first glance impossible, but is however feasible following an idea of Hambly (Figs. 2.21 and 2.22).

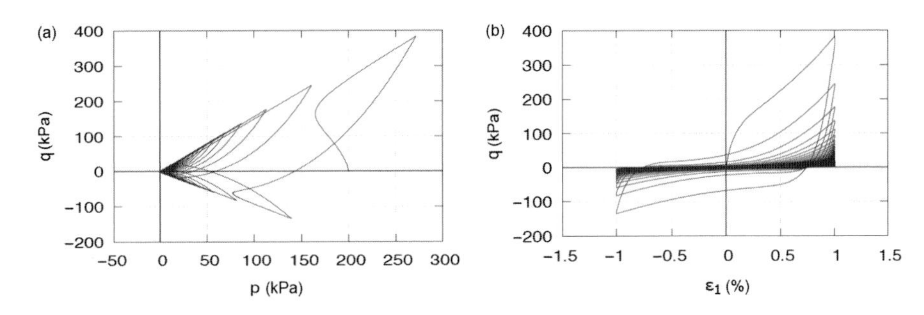

Figure 2.18 Undrained triaxial tests with sand by Wichtmann [115]. Cyclic strain. (a) Stress path and (b) stress–strain curve. The stress path exhibits the typical butterfly pattern.

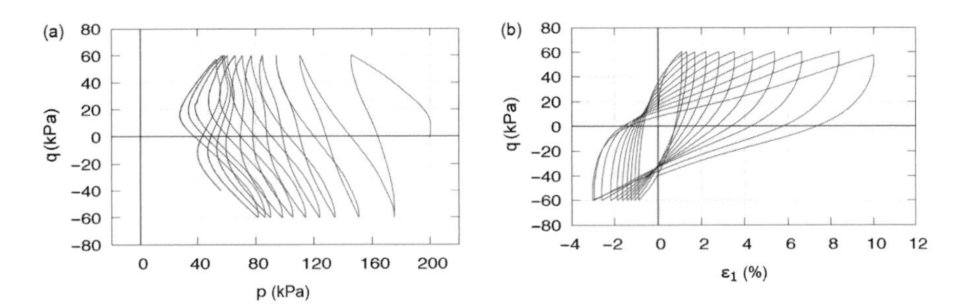

Figure 2.19 Undrained triaxial tests with clay by Wichtmann [115]. Cyclic stress. (a) Stress path and (b) stress–strain curve. The stress path exhibits the typical butterfly pattern.

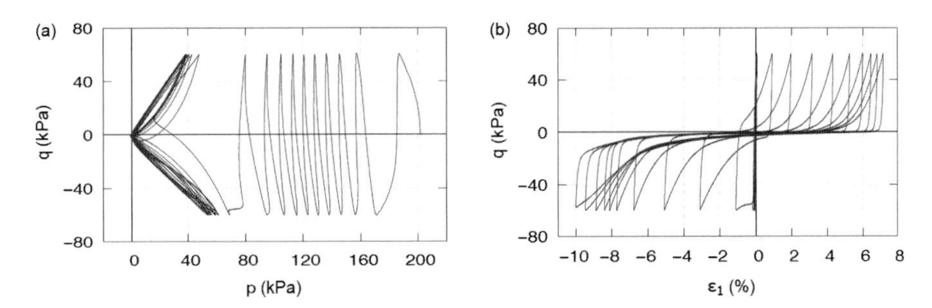

Figure 2.20 Undrained triaxial tests with sand by Wichtmann [115]. Cyclic stress. (a) Stress path and (b) stress–strain curve. The stress path exhibits a typical butterfly pattern.

Based on his experiments with sand in a true triaxial apparatus, Goldscheider [30] formulated two general rules:

2.8.1 The Rules of Goldscheider:

1. The response to a proportional strain path with a stress-free initial state, and only to such a strain path, is a proportional stress path.
2. In the case of a proportional strain path with non-zero stress in the initial state, the stress tends on a curve in the stress space asymptotically towards that straight line through the zero-stress point which would occur as a stress path in the case of a proportional strain path with a stress-free initial state.

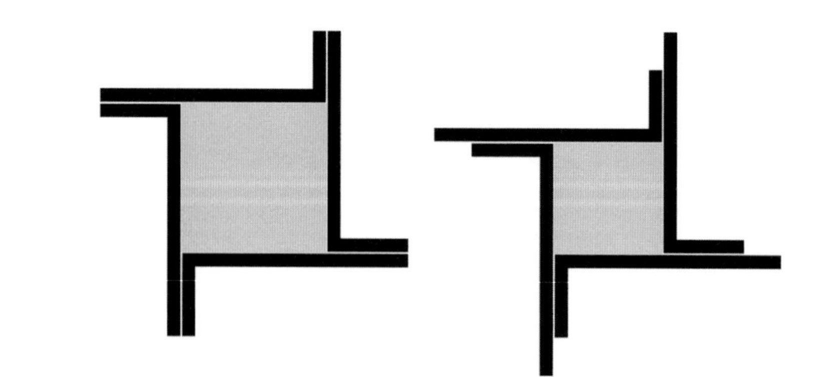

Figure 2.21 Rectilinear extension according to the principle of Hambly.

Figure 2.22 Biaxial apparatus for plane deformation of clay samples according to the principle of Hambly [62, 107].

The soil sample is cuboidal and experiences kinematic boundary conditions from all sides. One would therefore expect the deformation to be homogeneous. However, this is not always the case. Kuntsche [62], Topolnicki [107] as well as Desrues et al. [18] report of shear bands in clay samples (Fig. 17.5).

2.9 Simple Shear

The confinement of the sample imposes simple shear, see Fig. 2.23. The wall friction cannot be adjusted to equal the internal friction acting within the sample. Thus, the

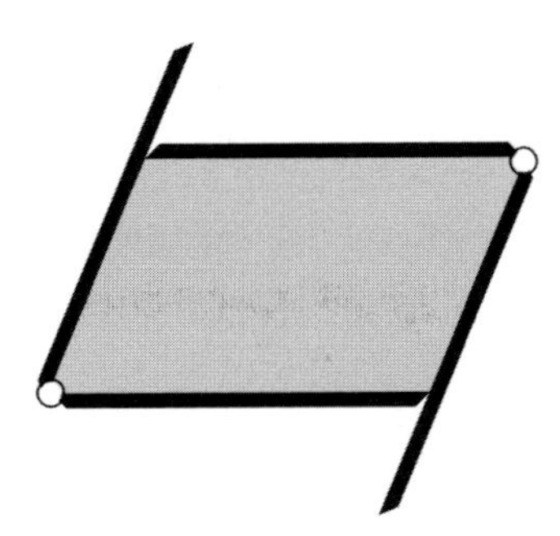

Figure 2.23 Simple shear apparatus, schematically.

stress state within the sample proves to be inhomogeneous [10] and, hence, simple shear tests are rather rare. Their results have a marked similarity to the results of triaxial tests, given the correspondence between the shear stress τ and the stress deviator $(\sigma_1 - \sigma_2)/2$.

2.10 Strain- versus Stress-Control

Deformation of samples can be either *stress-controlled* or *strain-controlled*. This means that either the stress is applied and the deformation is monitored, or vice versa. With strain-contolled tests, the deformation imposed to a sample by a piston proceeds with constant velocity v. With stress-controlled tests, the loads are (manually) applied in steps. At first glance, both conditions appear equivalent, since cause and effect are equivalent in physics. However, in both cases caveats with respect to the results (i.e. stress–strain curves and stress paths) must be taken into account:

- does the velocity v play a role? This is the case if the material is *rate sensitive* or *rate dependent*. Soils are in fact rate dependent: changes of velocity induce changes of stress. In the ideally static case, the test should be carried out with zero velocity, which is of course a contradiction.
- does the length of the time intervals between the individual load steps play a role? This is in fact the case in view of creep, i.e. the increase of strain with time at constant stress. Soils exhibit creep, and this fact makes stress strain measurements more or less questionable. They are only valid under the assumption that the time intervals at each load step are 'sufficiently' long so that constant strains are attained.

There exists some evidence that the results of strain-controlled and stress-controlled tests do not coincide [38].

2.11 Role of Time

Time-dependent effects like creep, relaxation, viscosity and ageing are additional aspects of the mechanical behaviour. The relevant notions are:

Creep: Increase of deformation at constant stress (Fig. 2.24). This effect is particularly disturbing, e.g. in stress-controlled oedometric tests. The settlement of the sample increases after the application of the load. This makes it impossible to clearly assign a deformation to a stress. Although the rate of deformation is getting smaller and smaller, it is not possible to extrapolate towards a final settlement. Creep is more pronounced with clay than with sand. In many cases it is neglected, in some other cases predefined deformations or creep rates are awaited.

Relaxation: Decay of stress at fixed deformation (Fig. 2.24). Relaxation can be observed in strain-controlled tests. When the motion of the loading piston is stopped, a gradual drop in stress is registered, which is usually proportional to the logarithm of time: $\Delta\sigma \propto \ln t$. When driving is resumed, the stress rises quickly to its previous value. Hence, relaxation has a rather episodic character. Usually, the end of relaxation cannot be awaited, as the logarithmic function is not bounded. This drawback is often neglected, as relaxation soon becomes extremely slow.

Rate sensitivity: Rate sensitivity (or rate dependence) is a property according to which the rate of deformation plays a role, see also Section 18.4.3. One can strain a sample from $\varepsilon_1 = 0$ to $\varepsilon_1 = 5\%$ within an hour or a day. If the resulting stress is the same, then the material is rate independent. Rate sensitivity can be seen as a sort of viscosity and implies that the stress–strain curve of, e.g. triaxial tests, suffers a jump whenever the rate of deformation is abruptly changed, see Figs. 2.25 and 2.26.

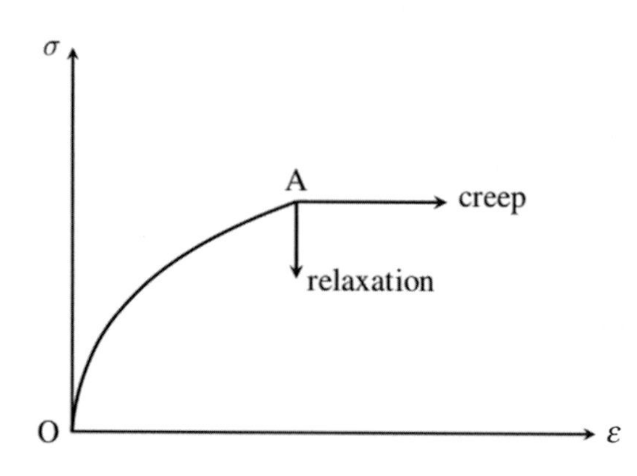

Figure 2.24 Creep and relaxation (schematically). Between points O and A the sample is deformed with constant deformation rate.

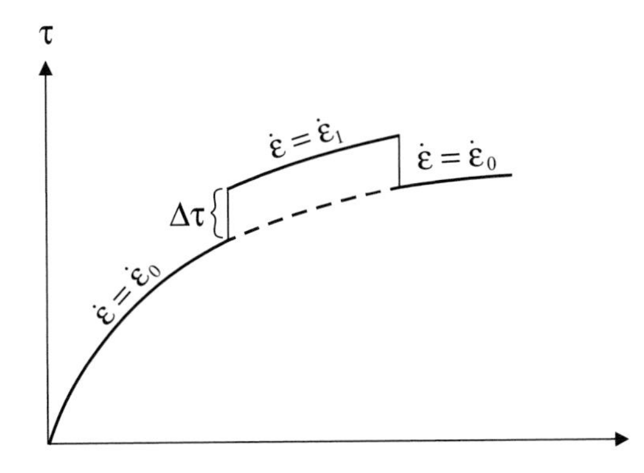

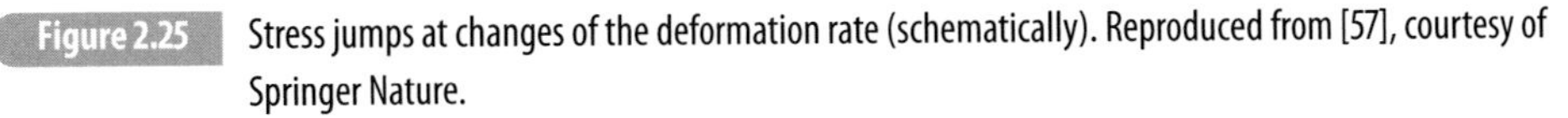

Figure 2.25 Stress jumps at changes of the deformation rate (schematically). Reproduced from [57], courtesy of Springer Nature.

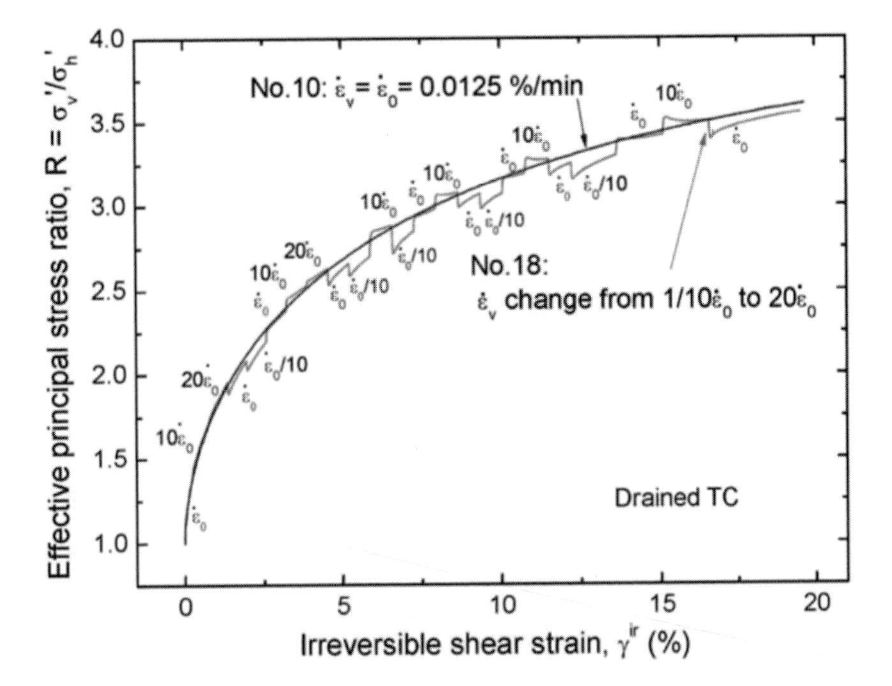

Figure 2.26 Triaxial tests with sand. Reproduced from [102]. σ_1/σ_3 is plotted vs $\varepsilon_1 - \varepsilon_3$. The reference test (thick line) is carried out with constant deformation rate. The thin line shows the results with jumps of the deformation rate and does not completely correspond to the pattern shown in Fig. 2.25.

Ageing: Stiffness and strength of soil increase with its age, i.e. the time lapse since its formation by sedimentation. This is manifested in the significant differences observed between undisturbed (natural) and remoulded (reconstituted) clay samples of equal void ratio. Some coarse-grained soils gradually attain a small cohesion which, however, is sensitive to minute deformations and usually disappears at sampling due to inevitable disturbances. This cohesion is attributed to tiny cementation bridges at the grain contacts.

2.12 Accuracy of Test Results

Laboratory tests are prone to various sources of error. The method of sample preparation may differ from laboratory to laboratory and has a large influence on the results, see Fig. 2.14. The density of samples exhibits spatial (and hardly detectable) variations. Creep and relaxation may have a considerable effect on the results. Friction of the pistons on their guide can falsify force measurements. Also, friction of the soil samples against their containing walls may produce errors. Last but not least, the onset of inhomogeneous deformation may have a more or less severe influence on the results. There are only a few round robin laboratory tests in soil mechanics. They indicate a considerable scatter of results across various laboratories. Consequently, the uncertainty of measurements is still unknown.

2.12.1 Some Statistical Aspects Related to Accuracy

Scatter implies that the results of measurements are stochastic (i.e. random) variables; unpredictable and yet subject to certain rules. The arithmetic mean of n measurements is $\bar{x} := (\sum x_i)/n$, and the 'empirical standard deviation' s is defined as:

$$s^2 := \frac{1}{n-1} \sum_n (\bar{x} - x_i)^2.$$

The uncertainty of a measurement series is $u = s/\sqrt{n}$. This value can be considered as the mean error of $\bar{x}$ while s can be considered as the mean error of a single measurement. A measured value should be indicated by $\bar{x} \pm u$. With regard to a measured value, one should only speak of the uncertainty and not of the accuracy. The complete indication of a measured value should also include the number n of measurements. If several measurements are available ($n > 1$), one must distinguish between the following cases:

1. **Repeatability within a laboratory:** the same observer, following a specified measurement procedure on the same object under the same test conditions (same measuring device) measures several times in short time intervals.
2. **Reproducibility across laboratories:** different observers carry out measurements on the same test object under different measuring instruments at different times.

Obviously, the standard deviation is larger for case 2 than for case 1. The repeatability standard deviation of a measuring device is a measure of its precision.

Functions $y(x_1, x_2, \ldots)$ of measured values $x_1, x_2, \ldots$ are likewise stochastic variables (random variables) in the sense that they are also subject to uncertainties. The question is how large the uncertainty of $y(x)$ is. The error propagation law of Gauss gives the standard deviation s_y of y as:

$$s_y = \sqrt{\sum \left(\frac{\partial y}{\partial x_i} s_i \right)^2} . \tag{2.3}$$

For the special case $y = x_1 + x_2 + \cdots + x_n$ with $s_{x_1} = s_{x_2} = \cdots = s$ we obtain $s_y = \sqrt{n}s$.

Remark In order to determine the height of Cheop's pyramide in Napoleon's Egyptian campaign, French officers measured the average height h of the 203 steps and its mean error m_h. The total height of the pyramid they then calculated to be $H = 203(h \pm m_h)$; but Fourier, who was present, pointed out that the height was correctly given by the quantity $H = 203h \pm \sqrt{203}\, m_h$ [98].

Range, **resolution** and **accuracy** are features of measuring instruments. Resolution is the smallest displayed change in the measured quantity. Usually, the resolution is much better and should not be confused with the measurement uncertainty. Accuracy is the ratio of the measurement uncertainty (measurement error) to the range, i.e. the largest measurable value of a measuring device.

2.13 Looking into the Samples

The soil samples tested in the laboratory are intransparent and, thus, only a visual inspection of their surface is possible. As long as their deformation is homogeneous (i.e. spatially constant, also called affine deformation), we can infer that the deformation within the sample is the same as the one derived from the deformation of its surface. In all other cases, scientists want to know the deformation in the interior of the sample. Several methods have been applied to this end. The clay sample in Figure 9.5 was vertically cut after the test and photographed with polarised light. Another method is to provide interlayers of coloured sand. In plane deformation, a glass wall makes it possible to observe the coloured layers (Fig. 2.27). Otherwise, the sample can be cut after the test, provided it has attained some cohesion (e.g. by moisture), and shear bands can be recognised by the offsets of the coloured sand layers. The displacement field can be obtained by means of 'digital image correlation' (DIC), also called 'particle image velocimetry', i.e. individual 'particles' are tracked by correlation of recognisable particle configuration patterns.

Again, another method is to use as granulate broken glass, which can be rendered transparent if the pores are filled with a fluid of the same refractive index. Then the stress trajectories can be visualised with polarised light (photoelasticity).

The penetration of X-rays has also been applied. The intensity of a radiation beam is diminished depending on the density of the passed distance. This effect, also used in medical radiography, enables us to detect shear bands, which have a higher porosity due to dilatancy. The displacement field can also be obtained by the images of embedded lead shots. By repeated radiation of stepwise rotated samples, a 3D image of the sample interior can be obtained (computed tomography, CT) [14]. The radiation is applied with X-rays, gamma rays, protons and neutrons. Synchrotrons

Figure 2.27 Shear band appearing in the case of active earth pressure. Courtesy Professor I. Herle, Institut für Geotechnik, Technische Universität Dresden.

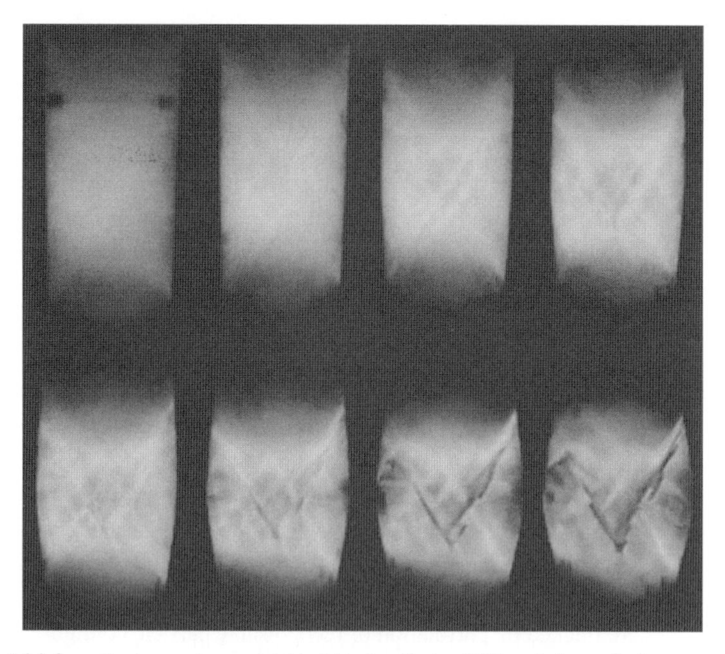

Figure 2.28 Incremental deformation in an axisymmetric triaxial test on Hostun HN31 sand observed using microtomography and spatial digital image correlation. Reproduced from [19].

as radiation sources provide a higher brilliance, which can also be achieved by laboratory-based micro-CT systems, however requiring increased measuring time. Computed tomography allows the application of digital image correlation also in three dimensions, see Fig. 2.28.

3 Mechanical Behaviour of Soil: Intuitively

3.1 Equations versus Intuition

Do we need equations to understand the behaviour of soil? It is often said that a complex behaviour can only be understood on the basis of a concise, constitutive equation. This is true to the extent that the equations describe the principle that underlies the complex behaviour. If this principle is understood, then one can qualitatively predict the behaviour even without equations. The equations describe the principle and offer some guidance to obtain quantitative predictions.

3.2 Proportional Paths for Granular Materials

As shown in Chapter 2, laboratory experiments reveal the extraordinary complexity of the mechanical behaviour of soil. Interestingly however, this behaviour can be, more or less, predicted with mere reasoning. To show this, we consider plane deformation ($\varepsilon_3 = 0$) so that the variables $\varepsilon_1, \varepsilon_2$ and σ_1, σ_2 are sufficient to describe several loading processes.

Proportional strain paths are characterised by $\varepsilon_1{:}\varepsilon_2$=const In terms of geometry, they are straight lines (rays) in the ε_1–ε_2 space, emanating from the origin $\varepsilon_1 = \varepsilon_2 = 0$. In this section we denote compressive strains and stresses as positive. Rays with $\varepsilon_1 + \varepsilon_2 < 0$ correspond to volume reduction and rays with $\varepsilon_1 + \varepsilon_2 > 0$ correspond to volume increase. Clearly, volume-increasing proportional strain paths that start from the stress-free state $\sigma_1 = \sigma_2 = 0$ are not feasible, because volume increase would imply that the individual grains lose contact, stress would vanish, and we could no longer consider this soil as a solid. Hence, the line $\varepsilon_1 + \varepsilon_2 = 0$ separates the feasible from the non-feasible proportional strain paths.

3.3 Relation between Strain Paths and Stress Paths

The question arises now as to which stress paths are associated with proportional strain paths? Before proceeding further, we have to notice that in a cohesionless dry sand only compressive stresses, i.e. $\sigma_1 > 0$ and $\sigma_2 > 0$, are feasible. To start with, we consider the stress-free state as the initial point of proportional strain paths. Strictly speaking, this state can only be realised in a zero-gravity space station. Hence, we

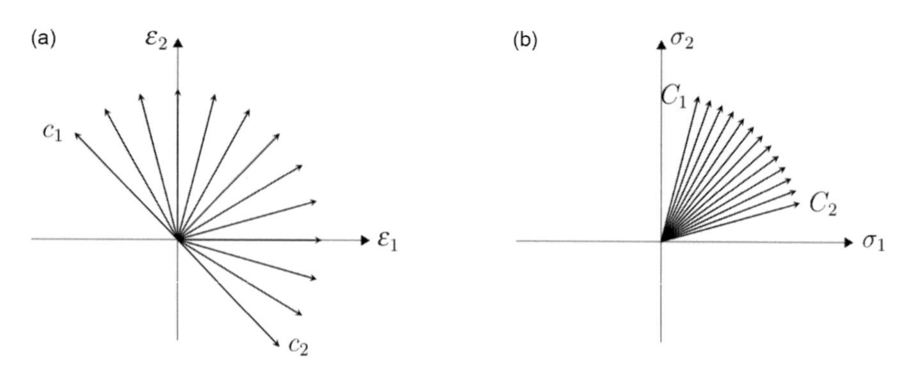

Figure 3.1 Proportional (a) strain paths and (b) corresponding stress paths for plane deformation ($\varepsilon_3 \equiv 0$). c_1 and c_2 are volume preserving strain paths, C_1 and C_2 are the corresponding proportional stress paths.

only assume $\sigma_1 \approx 0$ and $\sigma_2 \approx 0$ and we consider this a (practically) stress-free state. From this state only proportional strain paths with $\varepsilon_1 + \varepsilon_2 < 0$ can emanate. What do the corresponding stress paths look like? Are they curved or straight? At any rate, they have to remain within the quadrant $\sigma_1>0$, $\sigma_2>0$. Were they curved, which feature of the proportional strain paths would determine their curvature? Obviously, there is no such feature and, hence, it appears reasonable to assume that they are also straight, i.e. proportional paths. This is in fact what Goldscheider [30] found experimentally (Section 2.8.1).

Now, given the direction of a proportional strain path, what is the direction of the corresponding proportional stress path? Clearly, the stress path corresponding to $\varepsilon_1 = \varepsilon_2$ has to be $\sigma_1 = \sigma_2$ for an isotropic soil. But for $\varepsilon_1 \neq \varepsilon_2$ the directions of strain and stress paths cannot coincide because there are volume-reducing proportional strain paths outside the quadrant $\varepsilon_1>0$, $\varepsilon_2>0$, whereas the corresponding stress paths must lie within the quadrant $\sigma_1>0$, $\sigma_2>0$. To achieve this, the fan of proportional strain paths must be somehow narrowed to yield the fan of the corresponding stress paths. In particular, the two volume-preserving proportional strain paths $\varepsilon_1/\varepsilon_2 = -1$ (c_1 and c_2 in Fig. 3.1a) correspond to the proportional stress paths C_1 and C_2 that lie both in the first quadrant and delimit a particular region in the stress space (Fig. 3.1b). This is the locus $\mathcal{A}$ of stress states that are attainable if we start from the stress-free state and apply proportional straining.

3.4 Proportional Straining Starting at $\mathbf{T} \neq \mathbf{0}$

We now consider a non-vanishing stress state $\mathbf{T} \in \mathcal{A}, \mathbf{T} \neq \mathbf{0}$, and the fan of all proportional strain paths starting from there. Now volume *increasing* strain paths can also be applied. For a linear elastic material, the corresponding stress paths would also constitute a fan of straight lines pointing in all directions. In this way, tensile stress states could also be achieved that are not feasible for, say, dry sand (i.e. cohesionless soil). Thus, for dry sand the fan of stress paths associated with proportional strain paths starting from a state $\mathbf{T} \in \mathcal{A}, \mathbf{T} \neq \mathbf{0}$, must be somehow squeezed within the first quadrant. How can nature achieve this? One possibility is to choose a particular itinerary for each path. But nature 'prefers' general principles. A possible (and perhaps the only one) pertinent, general principle is composed from the following two statements:

- Volume-reducing proportional strain paths produce stress paths that asymptotically approach the proportional stress paths that would be obtained starting from the stress-free state. This means that the initial stress state is gradually forgotten and is known as the principle of fading memory, see the second rule of Goldscheider (Section 2.8.1).
- Volume-increasing proportional strain paths produce stress paths that strive to the stress-free state $\mathbf{T} = \mathbf{0}$. This is reasonable, as volume increase will sooner or later lead to loss of contacts between grains.

To conclude, let us take into account that the density ρ increases (or, equivalently, the void ratio e decreases) along volume-reducing proportional strain paths. What does this imply for the stress? With the mean pressure $p := (\sigma_1 + \sigma_2)/2$, we observe that the compressibilty $\mathrm{d}\rho/\mathrm{d}p$ is positive. This intuitive statement can also be inferred from thermodynamic stability. Taking into account that a dense soil is stiffer than a loose one we also infer that the compressibility decreases with increasing p. Thus, density and pressure increase along a proportional stress path within $\mathcal{A}$.

Taking these statements into account, the outcomes of laboratory tests with soil can be qualitatively predicted, as will be shown in the next sections. Thus, these statements express the constitutive law of soil. This will also be presented in the more precise language of equations using barodesy, as shown in Chapter 16.

3.5 Triaxial Tests

Let us consider the triaxial test in an ideal way, cleaned from all imperfections. A triaxial test starts with a consolidation, i.e. the soil sample undergoes first a proportional strain, say, a hydrostatic compression $\varepsilon_1 = \varepsilon_2 = \varepsilon_3$. Note that in the triaxial test we have $\varepsilon_2 \equiv \varepsilon_3$ throughout, hence, ε_3 is no more considered here. Subsequent to the consolidation phase, the triaxial test is conducted either in an undrained or drained way.

3.5.1 Undrained Triaxial Tests

With reference to a water-saturated soil sample, the word 'undrained' implies the condition of constant volume ('isochoric' deformation). This is the case if we assume that grains and water are incompressible so that changes of volume are only due to changes in the volume of the pores. According to Goldscheider's second rule, the stress path corresponding to an undrained or isochoric compression of the sample asymptotically approaches the proportional stress path C_1 of Figure 3.1. Of course, there are several modes of approach, the stress paths can turn either left or right, and this depends on the void ratio, i.e. on whether the sample is loose or dense, see Fig. 3.2. Clearly, a loose drained sample would contract whereas a dense drained

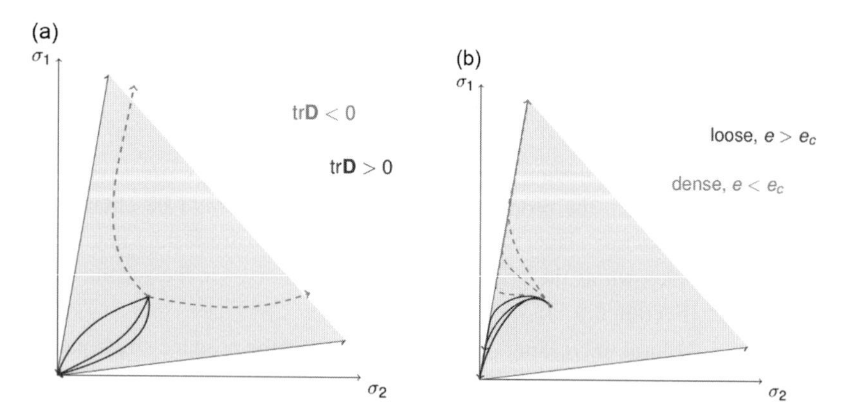

Figure 3.2 (a) Stress paths corresponding to proportional strain paths that do not start at $\mathbf{T} = \mathbf{0}$. (b) Stress paths corresponding to undrained triaxial tests.

sample would dilate. However, contractancy and dilatancy are inhibited in the here-considered undrained tests, and consequently a pore pressure builds up, which is higher in loose samples. This implies that the stress path of a loose sample turns more pronounced towards decreasing σ_2-values.

3.5.2 Drained Triaxial Tests

The conventional triaxial drained compression loading corresponds to a straight stress path σ_2 = const, see Fig. 3.3. As this path does not coincide with the curved stress paths pertinent to undrained tests, the volume (or the density or the void ratio) must change during this loading ('dilatancy'). The volume reduces or increases, depending on the initial void ratio of the sample. Clearly, this volume change must be limited, and this implies that eventually the deformation must become isochoric. This means that the stress state finally comes to the line C_1 of Fig. 3.1. On the other hand, it has to lie on the stress path σ_2 = const Hence, it must end at the intersection point of these two lines. In other words, the triaxial test leads finally to a state of constant stress and constant volume ('critical state'). If the initial density of the sample is high, the stress state initially overshoots the line C_1, and at a point ('peak') it necessarily turns back ('softening') and ends at C_1.

3.5.3 Cyclic Undrained Triaxial Tests

In the laboratory, drainage can be hindered by special precautions. Clearly, no such precautions can be realised in nature. And still, undrained deformation is very important in practice. Taking into account that drained deformation of water-saturated soil implies that water is squeezed out or sucked into the pores, and also that water has a viscosity, we see that drainage takes time. If the deformation is fast, as in the case of earthquakes, there is no time for drainage and the deformation is isochoric. Soil deformation due to an earthquake is not only undrained but also cyclic, i.e. it is composed of several cycles of loading and unloading.

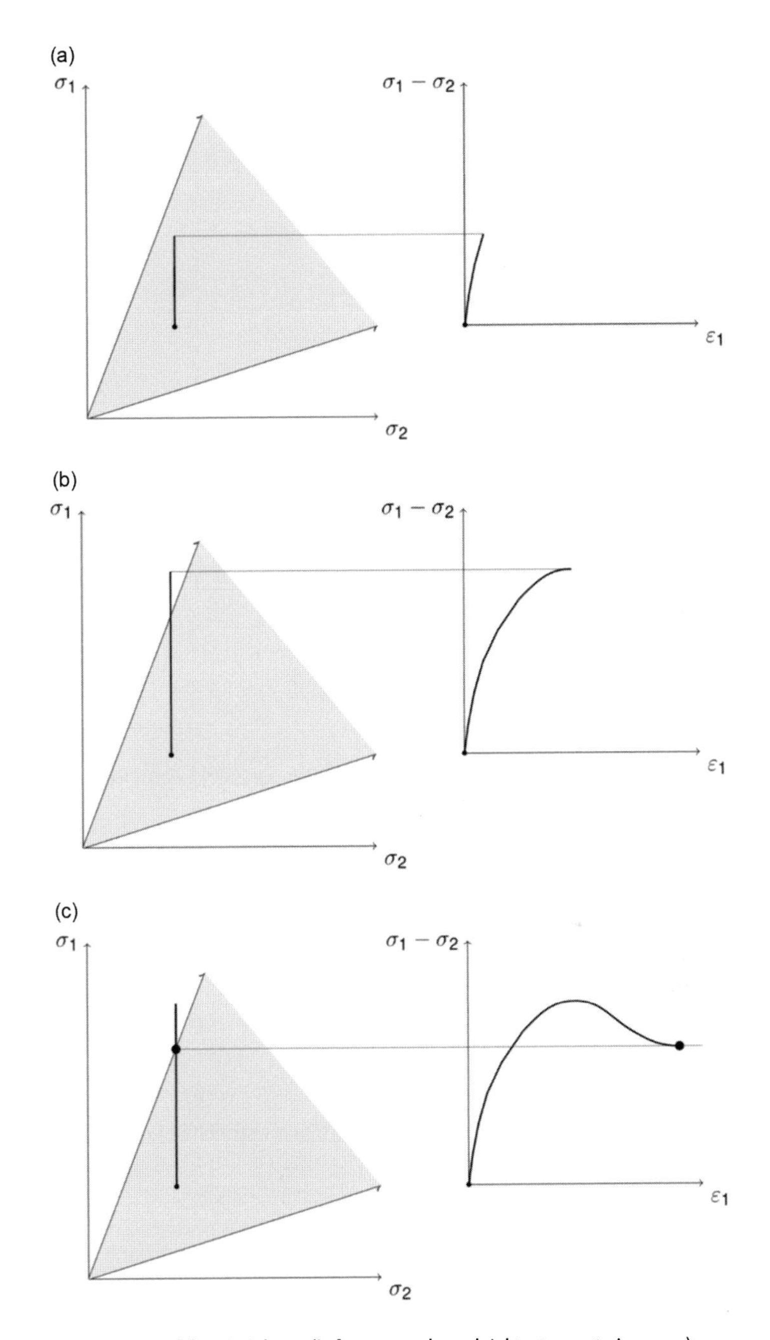

Three consecutive stages of the triaxial test. (Left: stress paths and right: stress–strain curves).

Figure 3.4 shows three consecutive stress paths for a loose (left) and a dense (right) sample. The stress paths can be inferred by reasoning according to the ideas stated above.

First loading: The stress paths caused by isochoric proportional straining must approach the path C_1 (Fig. 3.4a).

Unloading: Upon unloading the stress paths approach the extension limit stress state, i.e. the line C_2 (Fig. 3.4b).

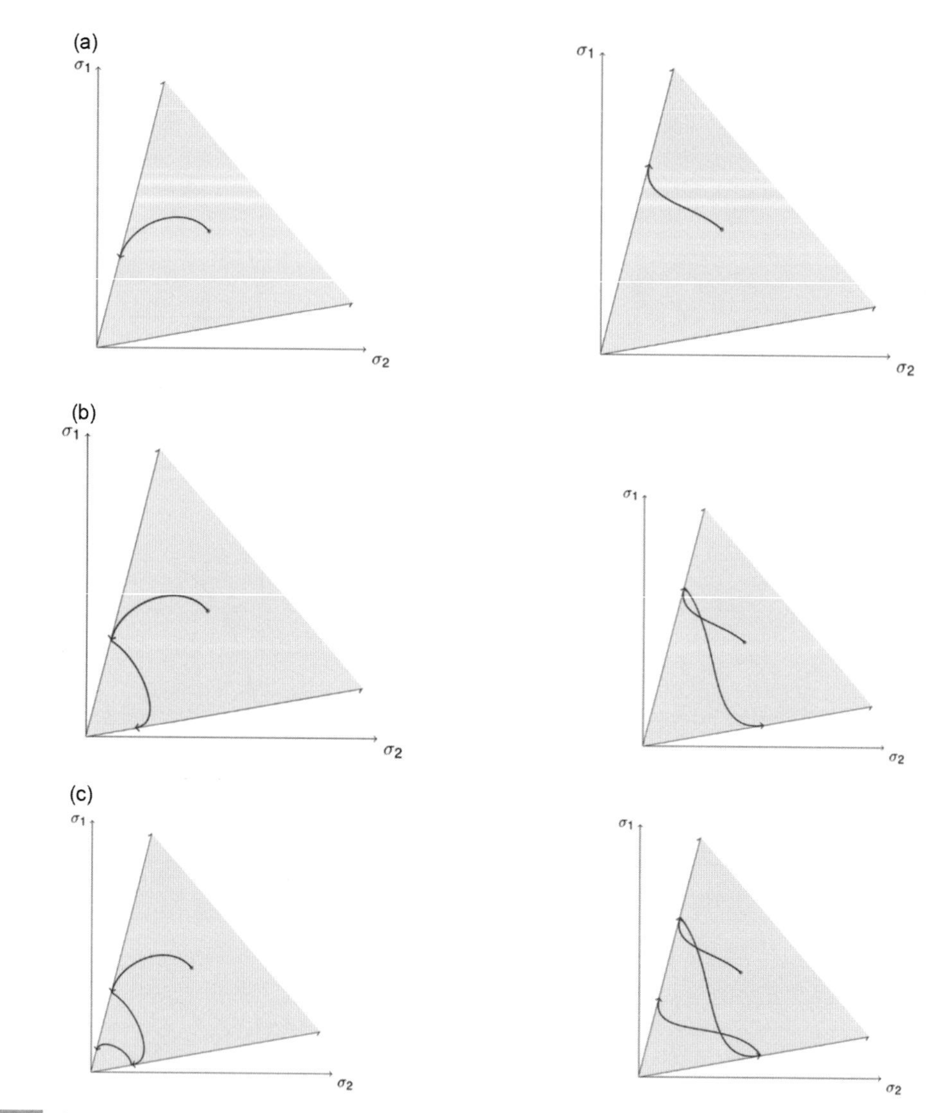

Figure 3.4 Consecutive stress paths at cyclic undrained triaxial tests with loose (left column) and medium dense (right column) sand.

Reloading: Upon reloading the stress paths approach again the compression limit stress state, i.e. the line C_1 (Fig. 3.4c).

In all three cases, the paths of the loose samples turn more pronouncedly to the direction of decreasing σ_2 for the reason stated above (Fig. 3.4, left column).

4 Vectors and Tensors

4.1 Purpose of This Chapter

Mechanics deals with scalar quantities such as density, temperature and energy but also with vectors such as force and velocity. Transformations of vectors require a third type of quantity, namely tensors. Vectors and tensors are often strange concepts to readers, therefore, a short introduction is presented here. It is rather sloppy and by no means a rigorous introduction to the vast field of linear algebra.

4.2 Vectors

Vectors comprise several components written in an array:

$$(5 \quad 3 \quad 2) \quad \text{or} \quad \begin{pmatrix} 5 \\ 3 \\ 2 \end{pmatrix}. \tag{4.1}$$

It is convenient to enumerate the components writing, e.g. $a_1 = 5, a_2 = 3, a_3 = 2$. The general reference to a component reads a_i, and this symbol can also be used to represent the entire vector. Note that the components of a vector always refer to a system of basis vectors that must be predefined. Often, the basis vectors point to the directions of a Cartesian system of coordinates x, y, z and are denoted as $\mathbf{e}_x$, $\mathbf{e}_y$, $\mathbf{e}_z$. It facilitates reading to denote vectors with boldface letters. The indices x, y, z are often replaced by the numbers 1, 2, 3. Thus, the vector $(5 \quad 3 \quad 2)$, which we now call $\mathbf{a}$, should be understood as

$$\mathbf{a} = a_1\, \mathbf{e}_1 + a_2\, \mathbf{e}_2 + a_3\, \mathbf{e}_3. \tag{4.2}$$

It is important to realise that one and the same vector can be represented by different components if we use different basis vectors. If, for example, we interchange the basis vectors $\mathbf{e}_1$ and $\mathbf{e}_2$, then $\mathbf{a}$ would read $\mathbf{a} = (3 \quad 5 \quad 2)$.

The scalar and the cross products of two vectors are:

$$\mathbf{a} \cdot \mathbf{b} = \begin{pmatrix} a_1 \\ a_2 \\ a_3 \end{pmatrix} \cdot \begin{pmatrix} b_1 \\ b_2 \\ b_3 \end{pmatrix} = a_1 b_1 + a_2 b_2 + a_3 b_3 = \sum_{i=1}^{n} a_i b_i \tag{4.3}$$

$$\mathbf{a} \times \mathbf{b} = \begin{pmatrix} a_2 b_3 - a_3 b_2 \\ a_3 b_1 - a_1 b_3 \\ a_1 b_2 - a_2 b_1 \end{pmatrix}. \tag{4.4}$$

Vectors with $\mathbf{a} \cdot \mathbf{b} = 0$ are *normal* (or 'orthogonal') to each other. The Euclidean norm (or 'length') of a vector $\mathbf{a}$ is

$$|\mathbf{a}| = \sqrt{\sum_{i=1}^{3} a_i a_i} = \sqrt{a_1 a_1 + a_2 a_2 + a_3 a_3}. \tag{4.5}$$

Usually, basis vectors have length 1 and are normal to each other. Attention should be paid to this fact, as some equations are valid only for this case. Vectors can have dimensions of 2, 3 and so on.

We can consider the values of a function $f(x)$ in the range $a \le x \le b$ at the equidistant points $x_0 = a, x_1 = a + h, x_2 = a + 2h, x_3 = a + 3h, \ldots, x_n = b$ with $h = (b - a)/n$. The values $f(x_0), f(x_1), f(x_2), \ldots$ constitute a vector of dimension n. For $n \to \infty$ we can consider the function $f(x)$ itself as a vector in a so-called Hilbert space. The scalar product of the vectors $f(x)$ and $g(x)$ is defined as $\int_a^b f(x)g(x)\mathrm{d}x$.

4.3 Tensors

The transformation of vectors is important in continuum mechanics. In analysing deformation, we want to know how an infinitesimal material vector $d\mathbf{X}$ is changed (transformed) to $d\mathbf{x}$ by deformation. In analysing stress, we want to know the relation between the unit normal vector $\mathbf{n}$ upon an infinitesimal surface element dS and the stress vector $\mathbf{t}$ (= force per unit area) acting thereupon. It turns that this transformation is linear: $\mathbf{t} = \mathbf{T}\mathbf{n}$ or

$$t_1 = T_{11}n_1 + T_{12}n_2 + T_{13}n_3 \tag{4.6}$$

$$t_2 = T_{21}n_1 + T_{22}n_2 + T_{23}n_3 \tag{4.7}$$

$$t_3 = T_{31}n_1 + T_{32}n_2 + T_{33}n_3, \tag{4.8}$$

(4.9)

which can also be written as

$$t_i = \sum_{j=1}^{3} T_{ij} n_j. \tag{4.10}$$

The array of the coefficients

$$\begin{pmatrix} T_{11} & T_{12} & T_{13} \\ T_{21} & T_{22} & T_{23} \\ T_{31} & T_{32} & T_{33} \end{pmatrix}$$

is called a matrix. Often, the words 'tensor' and 'matrix' are interchangeable.

Considering Equations (4.3), (4.5) and (4.10), we observe that the summation $\sum$ is carried out over the index that appears twice in the following expression. It is, therefore, practical to execute the summation without writing the symbol $\sum$. It suffices to see that an index appears twice:

$$t_i = \sum_{j=1}^{3} T_{ij} n_j = T_{ij} n_j. \tag{4.11}$$

This so-called summation convention proves convenient as soon as one gets (with some practice) familiar with it. We will use it henceforth.

With $y_i = A_{ij}x_j$ and $z_k = B_{ki}y_i$ we can write $z_k = B_{ki}A_{ij}x_j$ or $z_k = C_{kj}x_j$. Herein, C_{kj} is the product of B_{ki} and A_{ij}:

$$C_{kj} = B_{ki}A_{ij}. \tag{4.12}$$

Again, $B_{ki}A_{ij}$ is an abbreviation for $\sum_{i=1}^{3} B_{ki}A_{ij}$. To concentrate on the physical meaning of mathematical relations, we often omit the indices and denote vectors and tensors simply by using boldface letters. Thus, **A** is the so-called symbolic notation for A_{ij} and **v** the one for v_i.

Analogous to the Euclidean norm for vectors, we have the Frobenius norm for a tensor:

$$|A_{ij}| = \sqrt{A_{ij}A_{ji}}. \tag{4.13}$$

The transpose of A_{ij} is defined as $A_{ij}^T = A_{ji}$. The inverse of **A** is defined by the equation $\mathbf{A}\mathbf{A}^{-1} = \mathbf{1}$, where **1** is the unit tensor

$$\mathbf{1} = \begin{pmatrix} 1 & 0 & 0 \\ 0 & 1 & 0 \\ 0 & 0 & 1 \end{pmatrix}.$$

Its components are denoted by the Kronecker symbol

$$\delta_{ij} = \begin{cases} 1 & \text{for} \quad i = j \\ 0 & \text{for} \quad i \neq j. \end{cases} \tag{4.14}$$

A tensor **A** can be decomposed into symmetric and antisymmetric (or 'antimetric') parts: $\mathbf{A} = \mathbf{A}^S + \mathbf{A}^A$ with $\mathbf{A}^S = (\mathbf{A} + \mathbf{A}^T)/2$ and $\mathbf{A}^A = (\mathbf{A} - \mathbf{A}^T)/2$.

The dyadic product of two vectors is a tensor defined as follows:

$$(\mathbf{x} \otimes \mathbf{y})\,\mathbf{z} = \mathbf{x}\,(\mathbf{y} \cdot \mathbf{z}),$$

where $(\mathbf{y} \cdot \mathbf{z})$ is the scalar product of **y** and **z**.

We can associate to the vector F^k a tensor F_{ij} via the equation $F_{ij} = e_{ijk}F^k$. e_{ijk} is the so-called permutation symbol defined as

$$e_{123} = e_{312} = e_{231} = 1, \quad e_{132} = e_{213} = e_{321} = -1, \quad \text{else:} \quad e_{ijk} = 0. \tag{4.15}$$

From this equation results, in particular: $e_{mlk}e_{mlk} = 2$. The cross product $\mathbf{c} = \mathbf{a} \times \mathbf{b}$ can be written as:

$$c_i = e_{ijk}a_jb_k.$$

4.3.1 Invariants

As stated, the components of the vectors connected via a tensor, e.g. $\mathbf{t} = \mathbf{T}\mathbf{n}$, depend on the used coordinates and the related basis vectors, which may be different for each of the two vectors. Correspondingly, the components of a tensor depend on the underlying basis vectors. Thus, one and the same physical entity, such as the stress tensor, can be represented by different components. This situation is similar with a verbal statement, which has different representations depending on the language used to express it. Interestingly, we can assign to a tensor numbers, the so-called invariants, that do not change when the basis vectors are changed. One invariant is the so-called trace of a tensor

$$\mathrm{tr}\mathbf{A} = A_{ii} = A_{11} + A_{22} + A_{33}. \tag{4.16}$$

Other invariants are $\mathrm{tr}\mathbf{A}^2$ and $\mathrm{tr}\mathbf{A}^3$, and of course any of their infinitely many combinations, e.g. $17\mathrm{tr}\mathbf{A}^2 - 5\mathrm{tr}\mathbf{A}^3$. The invariants alone do not suffice to represent a tensor, which also contains information on some directions, the so-called eigendirections. They result from the equation $\mathbf{An} = \lambda\mathbf{n}$.

5 Fields

5.1 Fields in Continuum Mechanics

Continuum mechanics overlooks the composition of materials from atoms, molecules and grains and assumes that mass is continuously distributed over a region. Hence, it assigns values to points $\mathbf{x}$ of space in a continuous way. Functions of spatial position $\mathbf{x}$, $y = f(\mathbf{x})$, are called fields. y can be a scalar quantity (e.g. temperature, density), a vector (e.g. velocity) or a tensor (e.g. stress, deformation). Field theory is mathematically demanding. Here, we will only consider some basic notions that are of importance in continuum mechanics.

5.2 Coordinates

Points of a three-dimensional space can be identified by three numbers, the coordinates x_1, x_2, x_3. It makes sense to assign the values of the coordinates in some ordered way such that we can imagine in the space threads, along which the coordinate x_i varies (so-called coordinate lines), and sheets, on which the coordinate x_i remains constant (so-called coordinate surfaces).

To treat vectors and tensors we need not only coordinates but also basis vectors. We can assign each point a basis vector (and also another two basis vectors that are perpendicular to each other) in two different ways: Either as parallel to the coordinate lines, $\mathbf{e}_i$, or as normal to the coordinate surfaces, $\mathbf{e}^i$. These two basis vectors coincide in the special case that the coordinates are orthogonal. But even orthogonal coordinate lines can be curvilinear. In the special case that the coordinate lines are straight lines and orthogonal to the coordinate surfaces we have a so-called Cartesian system of coordinates.

Clearly, Cartesian systems are the simplest ones. However, there are cases where other systems are advantageous, e.g. for axisymmetric problems, such as a pile or a shaft, the equations become considerably simpler if we use the so-called cylindrical coordinates r, θ, z instead of the Cartesian coordinates x, y, z. Cylindrical coordinates are orthogonal curvilinear coordinates; the lines r = const are circles. When using curvilinear coordinates, one must take into account that the basis vectors $\mathbf{e}_i$ can have different directions at different points. If we attach the coordinates to the points of a deformable continuum, i.e. if we identify coordinate lines with material lines, then an initially orthogonal system can become non-orthogonal after deformation.

The multiplicity of coordinate systems poses the task of how to transform entities that are expressed with a particular system of coordinates into another system.

5.3 Vector Fields

Velocity fields are vector fields, considered, e.g. in hydrodynamics and aerodynamics. Soil mechanics usually considers slow processes, e.g. the settlement of the underground and the deformation of a soil sample in a triaxial test. So, the involved velocities are small. Larger velocities appear during earthquakes, landslides, etc. Velocities $\mathbf{v}$ and displacements $\Delta\mathbf{u}$ are related by $\mathbf{v} \approx \Delta\mathbf{u}/\Delta t$.

A velocity field is given by three scalar equations that express the components of the velocity as functions of the position in space, i.e. $v_1(x_1, x_2, x_3)$, $v_2(x_1, x_2, x_3)$ and $v_3(x_1, x_2, x_3)$. Gradient fields, $\mathbf{v} = \operatorname{grad}\Phi = \nabla\Phi$ are vector fields that can be given by only one scalar function $\Phi(x_1, x_2, x_3)$.

As stated, the representation of vectors and tensors needs basis vectors. In curvilinear coordinate systems, the basis vectors vary from one point to another. This may lead to the fact that the components of, say, a constant velocity of a particle vary with the position of the particle.

A good visualisation of planar velocity fields is given by the streamlines, which are tangential to the velocity at each point. For steady motions (i.e. motions that do not change with time), streamlines consist always of the same particles.

The following quantities are connected to vector fields:

- the scalar div $\mathbf{v}$, also written as $\nabla\cdot\mathbf{v}$. In Cartesian coordinates it is defined as $\operatorname{div}\mathbf{v} = \frac{\partial v_1}{\partial x_1} + \frac{\partial v_2}{\partial x_2} + \frac{\partial v_3}{\partial x_3}$, and in cylindrical coordinates: $\operatorname{div}\mathbf{v} = \frac{\partial v_r}{\partial r} + \frac{v_r}{r} + \frac{1}{r}\frac{\partial v_\theta}{\partial \theta} + \frac{\partial v_z}{\partial z}$. Isochoric velocity fields preserve the volume and are characterised by div $\mathbf{v} = 0$.
- the vector rot $\mathbf{v}$, also written as curl$\mathbf{v}$ and $\nabla\times\mathbf{v}$, reads in cartesian and cylindrical coordinates:

$$\operatorname{rot}\mathbf{v} = \begin{pmatrix} \frac{\partial v_3}{\partial x_2} - \frac{\partial v_2}{\partial x_3} \\ \frac{\partial v_1}{\partial x_3} - \frac{\partial v_3}{\partial x_1} \\ \frac{\partial v_2}{\partial x_1} - \frac{\partial v_1}{\partial x_2} \end{pmatrix}, \quad \operatorname{rot}\mathbf{v} = \begin{pmatrix} \frac{1}{r}\frac{\partial v_z}{\partial \theta} - \frac{\partial v_\theta}{\partial z} \\ \frac{\partial v_r}{\partial z} - \frac{\partial v_z}{\partial r} \\ \frac{\partial (r v_\theta)}{\partial r} - \frac{\partial v_r}{\partial \theta} \end{pmatrix},$$

respectively.

The following identities are valid: $\operatorname{div}\operatorname{rot}\mathbf{x} \equiv 0$ and $\operatorname{rot}\operatorname{grad}\Phi \equiv \mathbf{0}$.

A vector field $\mathbf{u}$ can always be represented as the sum of the rotor of another vector field $\mathbf{x}$ and of the gradient of a scalar field Φ (fundamental theorem of vector analysis [97]):

$$\mathbf{u} \equiv \operatorname{rot}\mathbf{x} + \operatorname{grad}\Phi. \tag{5.1}$$

5.3.1 Conformal Mapping

Potentials Φ for isochoric velocity fields, $\operatorname{div}\mathbf{v} = 0$, fulfil the Laplace equation $\operatorname{div}\operatorname{grad}\Phi = \Delta\Phi = 0$ (note that the expressions $\operatorname{grad}\Phi$ and $\nabla\Phi$, as well as $\operatorname{div}\mathbf{v}$ and $\nabla\cdot\mathbf{v}$, are equivalent). For plane isochoric velocity fields, i.e. for fields where the velocity $\mathbf{v}$ depends on x and y but not on z, we can easily find potentials Φ from analytical complex-valued functions $F(z)$, with $z = x + iy$ and $F = \Phi + i\Psi$. The

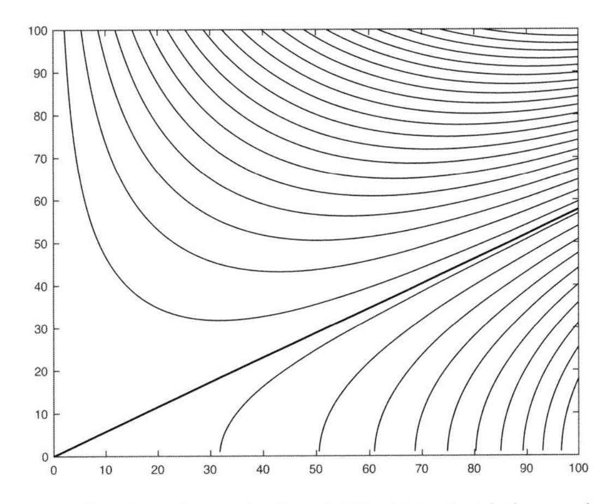

Figure 5.1 Streamlines between two walls intersecting each other at 60°, obtained with the complex-valued function $F(z) = z^3$.

functions $\Phi(x,y)$ and $\Psi(x,y)$ fulfil the LAPLACE equations $\Delta\Phi = 0$ and $\Delta\Psi = 0$. Curves $\Phi = \text{const}$ are (equi)potential lines. The lines $\Psi = \text{const}$ are orthogonal to them and represent thus streamlines. For instance, the function $z^n = r^n(\cos n\phi + i \sin n\phi)$ describes isochoric velocity fields of a flow enclosed by two lines with inclination of π/n to each other [87], see Fig. 5.1.

5.4 Continuous Fields and Discontinuities

The spatial distribution of a field quantity Ψ (such as density, velocity or stress) can be continuous or discontinuous [110]. In the latter case it undergoes a jump across a discontinuity, also called *singular surface* $\mathcal{S}$. Quantities Ψ on either side of a discontinuity are denoted by Ψ^+ and Ψ^- and the jump is denoted as $[\Psi] := \Psi^+ - \Psi^-$. Note that a quantity can be continuous while its spatial derivatives undergo a jump.

A singular surface can move. Using the unit normal vector $\mathbf{n}$, we obtain its normal velocity as $\mathbf{u} = u\mathbf{n}$. u is called *speed of displacement*.

A singular surface does not need to consist permanently of the same material points, i.e. it does not need to be a material surface. If a material point at the surface has the normal velocity $\dot{x}_n$, then the *speed of propagation* is a measure of the rate at which the moving surface traverses the material. More precisely, $U = u - \dot{x}_n$ is the normal speed of the surface with respect to the particles instantaneously situated upon it. A propagating singular surface is also called a *wave*.

With respect to jumps of velocity we distinguish between *vortex sheets* (the velocity suffers a transversal discontinuity) and *shock surfaces* (the normal velocity is discontinuous). Slip surfaces are vortex sheets.

Exercise 5.4.1 Determine the velocity field within a soil sample in an oedometer test.

Exercise 5.4.2 Determine the velocity field within a soil sample in an undrained triaxial compression test.

6 Deformation

6.1 Deformation and Grain Rearrangement

The deformation of granular materials consisting of uncrushable grains takes place as grain re-arrangement. This is a complex process of grains sliding and rotating relative to each other. The deformation of a continuum is completely different. In most cases, so-called topological deformation is considered, which implies that the individual material points pertain to their neighbours during deformation. Obviously, the continuum is a far-reaching idealisation of a granulate, which, however, proves very useful when we want to capture the observed stress–strain relationships by mathematical models.

6.2 How to Describe Deformation?

Deformation compares *two* configurations. Such comparisons are problematic. Compare, e.g. two pieces of, say, wood weighing 50 and 100 kg, respectively. Do they differ by 50% or 100% or 200%? Thus, there are many different mathematical expressions for deformation. Their differences are subtle and mirror the difficulty of the underlying concepts.

Remark Consider a one-dimensional example: We compress a sample of initially 10 cm in length to 9 cm (step A) and then to 8 cm in length (step B). If we use the 'engineering' measure of strain, $\varepsilon = \Delta l/l$, we obtain: $\varepsilon_A = -1/10 = -0.100$ and $\varepsilon_B = -1/9 = -0.111$. However, the sum of these strains, $\varepsilon_A + \varepsilon_B = -0.211$, is not equal to the strain corresponding to the compression 10 cm $\to$ 8 cm, namely, $\varepsilon_{A+B} = -2/10 = -0.200$. This inconsistency is removed if we use the so-called logarithmic strain $\epsilon = \ln(l/l_0)$, which yields the following values for our example: $\epsilon_A = \ln\frac{9}{10} = -0.105$; $\epsilon_B = \ln\frac{8}{9} = -0.118$; $\epsilon_A + \epsilon_B = -0.223$; $\epsilon_{AB} = \ln\frac{8}{10} = -0.223$.
For small strains, the difference between these two strain measures is small (Table 6.1).

6.3 Euler and Lagrange Approaches

A material point moves from its initial coordinates X_α $(\alpha = 1, 2, 3)$ to its actual coordinates x_i $(i = 1, 2, 3)$. We can consider the coordinates X_α as the names of the material points. These remain unchanged during a motion, which is given by

$$x_i = \chi_i(X_\alpha, t) \quad \text{or (in less precise notation)} \quad x_i = x_i(X_\alpha, t).$$

X_α are called material (or Lagrange) coordinates, whereas x_i are called spatial (or Euler) coordinates. We can trace a motion in two ways: either we consider a fixed material point X_α and ask about its spatial positions in time $x_i = x_i(X_\alpha, t)$ or we consider a fixed point x_i in space and ask by which material points X_α it is occupied in the course of time. We thus have:

Material (or Lagrange) description: $x_i = x_i(X_\alpha, t)$, field quantities are attached to X_α.

Spatial (or Euler) description: $X_\alpha = X_\alpha(x_i, t)$, field quantities are attached to x_i.

6.4 Deformation Gradient

Of central importance in analysing deformation is the *deformation gradient*

$$\mathbf{F} = F_{i\alpha} = x_{i,\alpha} = \partial x_i / \partial X_\alpha. \tag{6.1}$$

It is composed of a rotation $\mathbf{R}$ and a stretch tensor:

$$\mathbf{F} = \mathbf{RU} = \mathbf{VR}. \tag{6.2}$$

$\mathbf{U}$ is the 'right stretch tensor' and $\mathbf{V}$ is the 'left stretch tensor'. In general, $\mathbf{U} \neq \mathbf{V}$. The rotation tensor $\mathbf{R}$ is 'orthogonal', i.e. $\mathbf{R}^T = \mathbf{R}^{-1}$ or $\mathbf{RR}^T = \mathbf{1}$, hence

$$\mathbf{U}^2 = \mathbf{F}^T\mathbf{F}, \quad \mathbf{V}^2 = \mathbf{FF}^T, \quad \mathbf{V} = \mathbf{RUR}^T. \tag{6.3}$$

The following deformation tensors can be introduced as well:

- $\mathbf{C} := \mathbf{U}^2 = \mathbf{F}^T\mathbf{F}$: right Cauchy–Green deformation tensor
- $\mathbf{B} := \mathbf{V}^2 = \mathbf{FF}^T$: left Cauchy–Green deformation tensor. $\mathbf{B}$ and $\mathbf{C}$ are related: $\mathbf{B} = \mathbf{RCR}^T$.

Motions with $\mathbf{R} \equiv \mathbf{0}$ (e.g. the deformation in the triaxial and oedometer test) are called *rectilinear extensions*.

Table 6.1. $\epsilon = \ln(1 + \varepsilon)$

ε	ϵ
−5%	−5.1%
−10%	−10.5%
−20%	−22.3%

6.5 Rotation

A rotation is not a deformation. Together with translation, rotation is a so-called rigid body motion. Such motions are relative notions. We can only say, body A moves relative to body B or vice versa. Both statements are true. This is trivial but in the past it gave rise to big disputes, think of the trial of Galilei, who claimed that the Earth moves relatively to the Sun.

As rotations play an important role in analysing motions, some related essentials will be presented here.

6.5.1 Rotation of Cartesian Coordinates

In two dimensions, a rotation by angle φ can be conceived as multiplication of a complex number with $e^{i\varphi}$: $(x + iy)e^{i\varphi} = (x + iy)(\cos\varphi + i\sin\varphi) = (x\cos\varphi - y\sin\varphi) + i(x\sin\varphi + y\cos\varphi)$, hence a point $\mathbf{x}$ moves to $\mathbf{x}'$: $\mathbf{x}' = \mathbf{Q}\mathbf{x}$, with

$$\mathbf{Q} = \begin{pmatrix} \cos\varphi & -\sin\varphi \\ \sin\varphi & \cos\varphi \end{pmatrix}.$$

Let x_i, $i = 1, 2, 3$, be the coordinates of the initial system and x'_i be the ones of the rotated system. They are obtained from the initial coordinates by means of the transformation $x'_i = Q_{ij}x_j$ or $\mathbf{x}' = \mathbf{Q}\mathbf{x}$. The rotation matrix $\mathbf{Q}$ is 'orthogonal', i.e. it has the properties $\det\mathbf{Q} = 1$ and $\mathbf{Q}^T = \mathbf{Q}^{-1}$, and preserves right angles. The basis vectors of the rotated system are obtained with the same rotation: $\mathbf{e}'_i = \mathbf{Q}\mathbf{e}_j$ or $\mathbf{e}'_i = Q_{ij}\mathbf{e}_j$. Hence, the components of any vector $\mathbf{v} = v_j\mathbf{e}_j = v'_i\mathbf{e}'_i = v'_iQ_{ij}\mathbf{e}_j$ read in the rotated system $v'_i = Q^{-1}_{ij}v_j$, i.e. $\mathbf{v}' = \mathbf{Q}^{-1}\mathbf{v}$. We should distinguish between the following two cases:

- The system of coordinates rotates by $\mathbf{Q}$. The same vector $\mathbf{v}$ reads in the rotated system $\mathbf{v}' = \mathbf{Q}^{-1}\mathbf{v}$ ('passive rotation').
- The vector $\mathbf{v}$ rotates by $\mathbf{Q}$ and reads in the unrotated system $\mathbf{v}' = \mathbf{Q}\mathbf{v}$ ('active rotation').

The components of a tensor $\mathbf{T} = T_{ij}\mathbf{e}_i \otimes \mathbf{e}_j$ read in the rotated system: $T'_{kl}\mathbf{e}'_k \otimes \mathbf{e}'_l = T'_{kl}Q_{ki}\mathbf{e}_i \otimes Q_{lj}\mathbf{e}_j$, hence $T'_{kl} = Q^{-1}_{ki}T_{ij}Q^{-1}_{lj} = Q^T_{ki}T_{ij}Q_{jl}$, or simply $\mathbf{T}' = \mathbf{Q}^T\mathbf{T}\mathbf{Q}$ (passive rotation). Replacing $\mathbf{Q}$ with $\mathbf{Q}^{-1}$ we obtain the active rotation: $\mathbf{T}' = \mathbf{Q}\mathbf{T}\mathbf{Q}^T$.

Formula of Rodrigues: If $\mathbf{v}$ rotates by angle θ according to the right-hand rule with respect to the axis with direction $\mathbf{k}$, the rotated vector reads

$$\mathbf{v}' = \mathbf{v}\cos\theta + (\mathbf{k}\times\mathbf{v})\sin\theta + \mathbf{k}\,(\mathbf{k}\cdot\mathbf{v})(1 - \cos\theta). \tag{6.4}$$

Remark Fig. 6.1 shows rotations about the vertical z-axis. The x-axis of the underlying Cartesian coordinate system is perpendicular to the plane of the paper. The rotation $\mathbf{Q}$ reads:

$$\mathbf{Q} = \begin{pmatrix} 0 & -1 & 0 \\ 1 & 0 & 0 \\ 0 & 0 & 1 \end{pmatrix}. \tag{6.5}$$

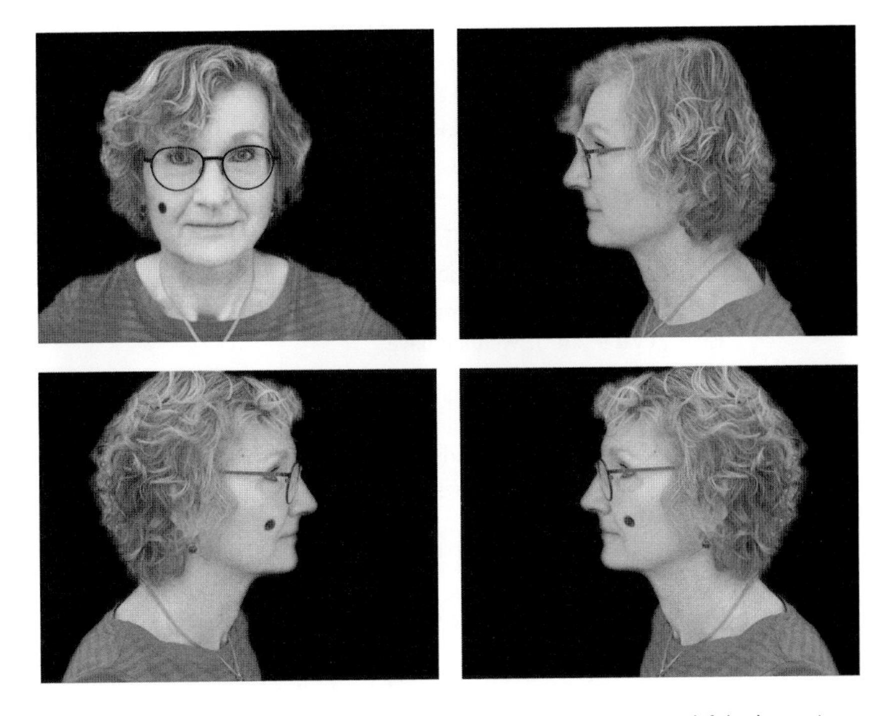

Figure 6.1 The top-right photograph results from the rotation $\mathbf{Q}$ of the person shown in the top-left (or the rotation $\mathbf{Q}^T$ of the observer). The bottom-left one results from the rotation $\mathbf{Q}^T$ of the person, whereas the bottom-right one results from the mirroring $\mathbf{S}$ of the bottom-left image.

The bottom-right figure results form the bottom-left one by the relation $\mathbf{x}' = \mathbf{S}\mathbf{x}$ with

$$\mathbf{S} = \begin{pmatrix} 1 & 0 & 0 \\ 0 & -1 & 0 \\ 0 & 0 & 1 \end{pmatrix}. \tag{6.6}$$

$\mathbf{S}$ mirrors the original image. Note that $\det\mathbf{S} = -1$, whereas $\det\mathbf{Q} = 1$.

6.6 Displacement Gradient

Considering the displacement $\mathbf{u}$ of a material point $\mathbf{X}$, we can write for a *Cartesian* coordinate system:

$$\mathbf{x} = \mathbf{X} + \mathbf{u}. \tag{6.7}$$

The restriction to Cartesian coordinates occurs because otherwise (i.e. if the base vectors are spatially variable) vectors may *not* be added or subtracted. We obtain

$$\mathbf{F} = \frac{\partial \mathbf{x}}{\partial \mathbf{X}} = \mathbf{1} + \frac{\partial \mathbf{u}}{\partial \mathbf{X}} = \mathbf{1} + \nabla\mathbf{u}, \tag{6.8}$$

hence

$$\mathbf{C} = \mathbf{U}^2 = \mathbf{F}^T\mathbf{F} = (\mathbf{1} + \nabla\mathbf{u}^T)(\mathbf{1} + \nabla\mathbf{u}) = \mathbf{1} + \nabla\mathbf{u} + \nabla\mathbf{u}^T + \nabla\mathbf{u}^T\nabla\mathbf{u}, \tag{6.9}$$

$$\mathbf{B} = \mathbf{V}^2 = \mathbf{F}\mathbf{F}^T = (\mathbf{1} + \nabla\mathbf{u})(\mathbf{1} + \nabla\mathbf{u}^T) = \mathbf{1} + \nabla\mathbf{u}^T + \nabla\mathbf{u} + \nabla\mathbf{u}\nabla\mathbf{u}^T. \tag{6.10}$$

6.6.1 Infinitesimal Strain

If we neglect the quadratic terms $\nabla\mathbf{u}\nabla\mathbf{u}^T$ and $\nabla\mathbf{u}^T\nabla\mathbf{u}$ in Equations 6.9 and 6.10 (geometric linearisation) we obtain $\mathbf{C} \approx \mathbf{B} \approx \mathbf{1} + 2\mathbf{E}$ with the 'infinitesimal strain' $\mathbf{E} := \frac{1}{2}(\nabla\mathbf{u} + \nabla\mathbf{u}^T)$. We then have

$$\mathbf{V} = \mathbf{B}^{1/2} \approx (\mathbf{1} + 2\mathbf{E})^{1/2} \quad \text{and} \quad \mathbf{U} = \mathbf{C}^{1/2} \approx (\mathbf{1} + 2\mathbf{E})^{1/2}.$$

Broken binomial series leads to $\mathbf{V} \approx \mathbf{U} \approx \mathbf{1} + \mathbf{E}$.
From $\mathbf{F} = \mathbf{R}\mathbf{U} = \mathbf{V}\mathbf{R}$ follows:

$$\mathbf{F} = \mathbf{1} + \nabla\mathbf{u} = \mathbf{R}\left[\mathbf{1} + \frac{1}{2}(\nabla\mathbf{u} + \nabla\mathbf{u}^T)\right] = \left[\mathbf{1} + \frac{1}{2}(\nabla\mathbf{u} + \nabla\mathbf{u}^T)\right]\mathbf{R}. \tag{6.11}$$

If quadratic terms are neglected, Equation 6.11 is fulfilled by

$$\mathbf{R} \approx \mathbf{1} + \frac{1}{2}(\nabla\mathbf{u} - \nabla\mathbf{u}^T). \tag{6.12}$$

With $\mathbf{V} \approx \mathbf{U} \approx \mathbf{1} + \frac{1}{2}(\nabla\mathbf{u} + \nabla\mathbf{u}^T)$ we obtain

$$\mathbf{U} + \mathbf{R} \approx \mathbf{V} + \mathbf{R} \approx 2 \cdot \mathbf{1} + \nabla\mathbf{u} \tag{6.13}$$

and finally:

$$\begin{aligned}\mathbf{F} &\approx \mathbf{U} + \mathbf{R} - \mathbf{1}\\ &\approx \mathbf{V} + \mathbf{R} - \mathbf{1}.\end{aligned}$$

The volumetric strain is $\mathrm{tr}\mathbf{E} = \frac{\Delta V}{V}$, the change of shape is given by the deviator $\mathbf{E}^* = \mathbf{E} - \frac{1}{3}\mathrm{tr}\mathbf{E}\,\mathbf{1}$.

Infinitesimal Strain in Cylindrical Coordinates

For some applications, the components of infinitesimal strain in cylindrical coordinates are useful. With the components u_r, u_θ, u_z of the displacement vector, they read:

$$\varepsilon_{rr} = \frac{\partial u_r}{\partial r}, \quad \varepsilon_{\theta\theta} = \frac{1}{r}\frac{\partial u_\theta}{\partial \theta} + \frac{u_r}{r}, \quad \varepsilon_{zz} = \frac{\partial u_z}{\partial z},$$

$$\varepsilon_{r\theta} = \varepsilon_{\theta r} = \frac{1}{2}\left(\frac{1}{r}\frac{\partial u_r}{\partial \theta} - \frac{u_\theta}{r} + \frac{\partial u_\theta}{\partial r}\right),$$

$$\varepsilon_{rz} = \varepsilon_{zr} = \frac{1}{2}\left(\frac{\partial u_r}{\partial z} + \frac{\partial u_z}{\partial r}\right), \quad \varepsilon_{\theta z} = \varepsilon_{z\theta} = \frac{1}{2}\left(\frac{1}{r}\frac{\partial u_z}{\partial \theta} + \frac{\partial u_\theta}{\partial z}\right).$$

For axisymmetric cases ($u_\theta = 0, \partial/\partial\theta = 0$) the non-vanishing components are:

$$\varepsilon_r = \frac{\partial u_r}{\partial r}, \quad \varepsilon_\theta = \frac{u_r}{r}, \quad \varepsilon_z = \frac{\partial u_z}{\partial z}.$$

6.7 Time and Spatial Derivatives

The time and the spatial derivatives of field quantities depend on whether the fields are spatial (Eulerian) or material (Lagrangean). With respect to velocity, it should be – strictly speaking – distinguished between

- spatial velocity $\mathbf{v}(\mathbf{x}, t)$ and
- material velocity $\dot{\mathbf{x}}(\mathbf{X}, t)$.

Table 6.2. Derivatives of spatial quantities Ω and of material quantities Φ

Spatial time derivative: $\Omega'(\mathbf{x}, t) = \frac{\partial \Omega(\mathbf{X},t)}{\partial t}$	Material time derivative: $\dot{\Phi}(\mathbf{X}, t) = \frac{\partial \Phi(\mathbf{X},t)}{\partial t}$
Spatial gradient: $\operatorname{grad} \Omega(\mathbf{x}, t) = \nabla_{\mathbf{x}} \Omega(\mathbf{x}, t)$	Material gradient: $\nabla \Phi(\mathbf{X}, t) = \nabla_{\mathbf{X}} \Phi$
Material time derivative of the spatial quantity $\Omega(\mathbf{x}, t)$: $\dot{\Omega}(\mathbf{x}, t) = \frac{d\Omega}{dt} = \frac{\partial \Omega}{\partial t} + \mathbf{v} \cdot \operatorname{grad} \Omega$	

6.7.1 Stretching

The symmetric part of the velocity gradient $\mathbf{L} = \operatorname{grad} \mathbf{v} = v_{i,j} = \dot{x}_{i,j}$ is the stretching tensor of Euler:

$$\mathbf{D} = D_{ij} = \frac{1}{2}(v_{i,j} + v_{j,i}) = \frac{1}{2}(\dot{x}_{i,j} + \dot{x}_{j,i}) = \dot{x}_{(i,j)}. \tag{6.14}$$

$\mathbf{D}$ is the time rate of the right stretch tensor $\mathbf{U}$ if the actual configuration is also the reference one:

$$\mathbf{D}(t) = \dot{\mathbf{U}}_{(t)}(t) = \frac{1}{2}\mathbf{R}(\dot{\mathbf{U}}\mathbf{U}^{-1} + \mathbf{U}^{-1}\dot{\mathbf{U}})\mathbf{R}^T. \tag{6.15}$$

For a small line element l with direction $\mathbf{n}$ we have:

$$\lim_{l \to 0} \frac{\dot{l}}{l} = D_{ij} n_i n_j. \tag{6.16}$$

$\mathbf{D}$ refers to the actual configuration and is *not* the time rate of any deformation measure for finite strain. Thus, the equations $\mathbf{D} = \dot{\mathbf{E}}$ and $\mathbf{E} = \int \mathbf{D} dt$ can be used in some cases but are not generally valid.

6.7.2 Spin

The spin tensor is the antisymmetric part of the velocity gradient: $\mathbf{W} = W_{ij} = \frac{1}{2}(v_{i,j} - v_{j,i}) = \frac{1}{2}\left(\dot{x}_{i,j} - \dot{x}_{j,i}\right) = \dot{x}_{[i,j]}$. Motions with $\mathbf{W} = \mathbf{0}$ are called irrotational. Caution: $\mathbf{W} \neq \dot{\mathbf{R}}$. Only when the reference configuration coincides with the current one may we write $\mathbf{W}(t) = \dot{\mathbf{R}}_{(t)}(t)$.

Irrotational motions deserve special attention, e.g. for them holds the theorem of Kelvin: If the normal velocity on the boundary of a body is given, the minimum kinetic energy is obtained for irrotational motions.

6.8 Example: Simple Shear

A dilatant shear is called simple shear (Fig. 6.2). A material point with initial coordinates (X_1, X_2) moves to (x_1, x_2):

$$x_1(t) = X_1 + X_2 \cdot f(t) \tag{6.17}$$

$$x_2(t) = X_2 + X_2 \cdot g(t). \tag{6.18}$$

The cases $g > 0$ and $g < 0$ correspond to dilatancy and contractancy, respectively. If the material and the spatial coordinates coincide for $t = 0$, we have $f(0) = g(0) = 0$. From Equations 6.17 and 6.18 follows:

$$\mathbf{F} = \frac{\partial \mathbf{x}}{\partial \mathbf{X}} = \begin{pmatrix} 1 & f \\ 0 & 1+g \end{pmatrix}.$$

The right and left Cauchy–Green tensors read:

$$\mathbf{C} = \mathbf{U}^2 = \mathbf{F}^T\mathbf{F} = \begin{pmatrix} 1 & f \\ f & f^2 + (1+g)^2 \end{pmatrix},$$

$$\mathbf{B} = \mathbf{V}^2 = \mathbf{F}\mathbf{F}^T = \begin{pmatrix} 1+f^2 & f(1+g) \\ f(1+g) & (1+g)^2 \end{pmatrix}.$$

The infinitesimal deformation tensor reads:

$$\mathbf{E} = \begin{pmatrix} 0 & f/2 \\ f/2 & g \end{pmatrix}.$$

The rotation tensor reads:

$$\mathbf{R} = \begin{pmatrix} R_{11} & R_{12} \\ R_{21} & R_{22} \end{pmatrix}.$$

It can be determined from $\mathbf{B} = \mathbf{R}\mathbf{C}\mathbf{R}^T$ or $\mathbf{R}^T\mathbf{B} = \mathbf{C}\mathbf{R}^T$, which yields four scalar equations for the components of

$$\mathbf{R} = \begin{pmatrix} \cos\alpha & -\sin\alpha \\ \sin\alpha & \cos\alpha \end{pmatrix}.$$

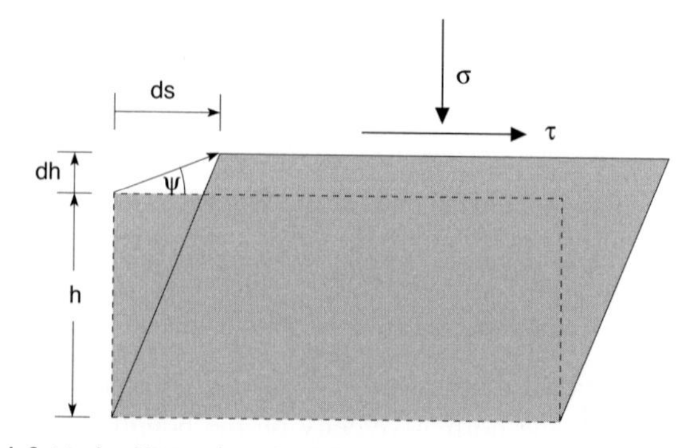

Figure 6.2 Simple shear is (by definition) a dilatant shear. Reproduced from [57], courtesy of Springer Nature.

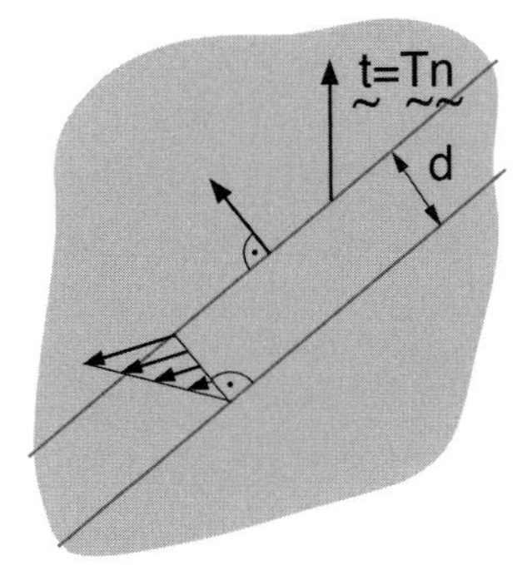

Figure 6.3 Simple shear in a shear band of thickness d. Reproduced from [57], courtesy of Springer Nature.

From the first equation we obtain:

$$\tan\alpha = \frac{-f}{2+g}.$$

The velocity gradient reads $\mathbf{L} = \operatorname{grad}\dot{\mathbf{x}} = \dfrac{\partial \dot{x}_i}{\partial x_j}$, with

$$\dot{x} = \begin{pmatrix} X_2\dot{f} \\ X_2\dot{g} \end{pmatrix} = \frac{1}{1+g}\begin{pmatrix} x_2\dot{f} \\ x_2\dot{g} \end{pmatrix}, \tag{6.19}$$

hence

$$\mathbf{D} = \frac{1}{2(1+g)}\begin{pmatrix} 0 & \dot{f} \\ \dot{f} & 2\dot{g} \end{pmatrix}, \quad \mathbf{W} = \frac{1}{2(1+g)}\begin{pmatrix} 0 & \dot{f} \\ -\dot{f} & 0 \end{pmatrix}. \tag{6.20}$$

Within a shear band of thickness d the motion is simple shear (Fig. 6.3). The linear distribution of velocity reads $\mathbf{v} = \frac{1}{d}(\mathbf{n}\cdot\mathbf{x})\mathbf{v}_0$. $\mathbf{n}$ is the unit vector normal to the shear band. With $\mathbf{v}_0(\mathbf{n}\cdot\mathbf{x}) \equiv (\mathbf{v}_0\otimes\mathbf{n})\mathbf{x}$ it follows $\mathbf{v} = \frac{1}{d}(\mathbf{v}_0\otimes\mathbf{n})\mathbf{x}$. Thus, the velocity gradient is obtained as:

$$\mathbf{L} = \frac{1}{d}(\mathbf{v}_0\otimes\mathbf{n}) \quad \rightsquigarrow \quad \mathbf{D} = \frac{1}{2d}(\mathbf{v}_0\otimes\mathbf{n} + \mathbf{n}\otimes\mathbf{v}_0). \tag{6.21}$$

6.9 Equations of Compatibility

Given a strain field $\varepsilon_{ij}(x_i)$, we ask whether the differential equation $\varepsilon_{ij} = \frac{1}{2}\left(u_{i,j} + u_{j,i} - u_{k,i}u_{k,j}\right)$ is integrable, so as to yield a continuous displacement field $u_i(x_j)$. Here, we use the abbreviation $u_{i,j} := \frac{\partial u_i}{\partial x_j}$. The equations of compatibility guarantee the integrability of this differential equation. If the displacement field is continuous, then the integral $\oint \varepsilon_{ij}\mathrm{d}s$ over a closed loop will vanish. Otherwise i.e. if the displacement field is not continuous (if the equations of compatibility are not fulfilled), then this integral yields a displacement. Regarding dislocations, this displacement is the so-called Burgers vector $\mathbf{b}$ (Fig. 6.4).

The violation of the equations of compatibility is the starting point of theories of continuously distributed dislocations and higher (or non-Boltzmann) continua. There, the individual material points undergo not only displacements but also

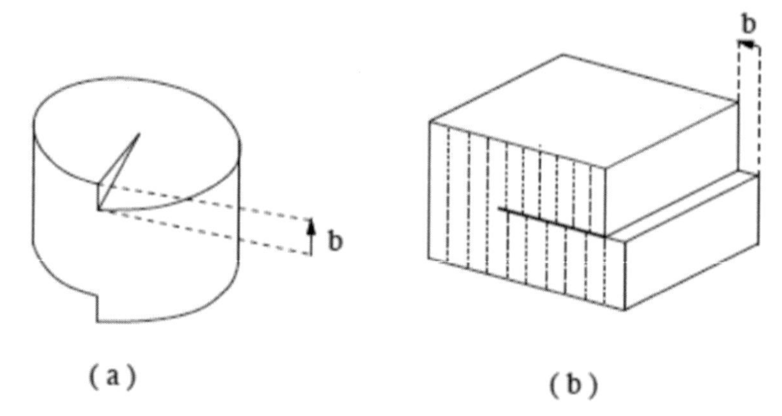

Figure 6.4 Screw (a) and edge (b) dislocations can be considered as 'atoms' of incompatible deformation. Reproduced from [46], courtesy of Springer Nature.

rotations (so-called Cosserat or oriented or polar continua) and even deformations. The rotation in a Cosserat continuum differs from the rotation **R**, as obtained from the deformation gradient (Equation 6.2) and is called incompatible rotation. The theories of higher continua are intricate and appealing (they even lead to spaces with non-Euclidean geometry). However, despite their theoretical appeal and many publications, they virtually contributed little to the mechanics of granular media.

The dislocation density α_{nk} is introduced by the equation:

$$db_k = \alpha_{nk} dF^n.$$

Herein b_k is the Burgers vector and dF^n an oriented surface element. The diagonal components of α_{nk} are *screw dislocations*, the other components are *edge dislocations* (Fig. 6.4).

7 Stress

7.1 What Is the Stress Tensor?

The stress tensor describes the forces that act upon surfaces within a continuum. The Cauchy stress tensor $\mathbf{T}$ is defined by the equation $\mathbf{t} = \mathbf{Tn}$. Herein, $\mathbf{n}$ is the unit normal vector on a (infinitesimal) cut surface and $\mathbf{t}$ is the stress vector (i.e. force divided by area) acting thereupon. Note that both, $\mathbf{t}$ and $\mathbf{n}$ are taken in the actual, i.e. the deformed configuration. There are also other definitions, e.g. $\mathbf{t} = \mathbf{T}_R\mathbf{n}_0$, where $\mathbf{T}_R$ is the so-called first Piola–Kirchhoff stress tensor (also called the Lagrange stress tensor) and $\mathbf{n}_0$ is the unit normal vector in the reference configuration. The transformation rule is $\mathbf{T} = \frac{1}{\det \mathbf{F}}\mathbf{T}_R\mathbf{F}^T$.

$\mathbf{T}$ is symmetric, i.e. $\mathbf{T} = \mathbf{T}^T$. This results from the balance law of moment of momentum and from the assumption that there are no 'couple stresses' (see Section 8.2.4).

The stress tensor can be decomposed of a deviatoric and a hydrostatic part:

$$\mathbf{T} = \mathbf{T}^* + \frac{1}{3}(\mathrm{tr}\mathbf{T})\mathbf{1}.$$

$\mathbf{T}^*$ is the deviator of $\mathbf{T}$. Clearly, $\mathrm{tr}\mathbf{T}^* = 0$.

The components of $\mathbf{T}$ are denoted either as T_{ij} or as σ_{ij} (Fig. 7.1):

$$\begin{pmatrix} \sigma_{11} & \sigma_{12} & \sigma_{13} \\ \sigma_{21} & \sigma_{22} & \sigma_{23} \\ \sigma_{31} & \sigma_{32} & \sigma_{33} \end{pmatrix}.$$

They depend on the system of coordinates. In principal axes they read

$$\text{either} \quad \begin{pmatrix} \sigma_1 & 0 & 0 \\ 0 & \sigma_2 & 0 \\ 0 & 0 & \sigma_3 \end{pmatrix} \quad \text{or} \quad \begin{pmatrix} T_1 & 0 & 0 \\ 0 & T_2 & 0 \\ 0 & 0 & T_3 \end{pmatrix}.$$

For compression, the values T_i are taken negative, whereas (following the tradition in soil mechanics) the values σ_i are taken positive. For axisymmetric problems (e.g. for the triaxial test) we have $\sigma_2 \equiv \sigma_3$, so the stress state can be fully expressed with only two variables, σ_1 and σ_2:

$$\mathbf{T} = \begin{pmatrix} \sigma_1 & 0 & 0 \\ 0 & \sigma_2 & 0 \\ 0 & 0 & \sigma_2 \end{pmatrix}, \quad \mathbf{T}^* = \frac{1}{3}\begin{pmatrix} 2(\sigma_1 - \sigma_2) & 0 & 0 \\ 0 & -(\sigma_1 - \sigma_2) & 0 \\ 0 & 0 & -(\sigma_1 - \sigma_2) \end{pmatrix}.$$

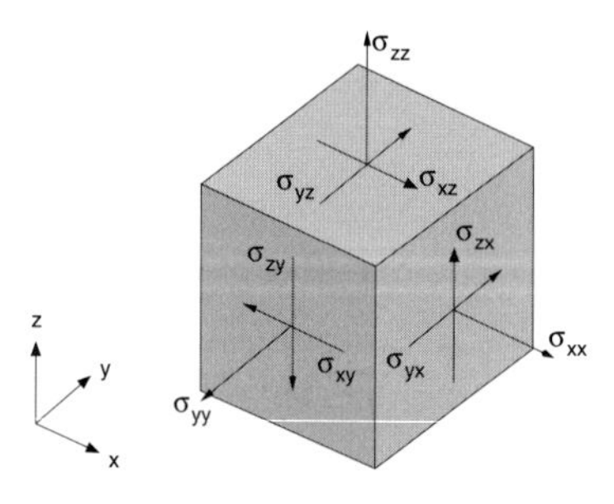

Figure 7.1 Stress components of a stress tensor in Cartesian coordinates. Its columns are the stress vectors acting upon planes $x_i = \text{const}, i = 1, 2, 3$, where $x_1 = x, x_2 = y, x_3 = z$. Reproduced from [57], courtesy of Springer Nature.

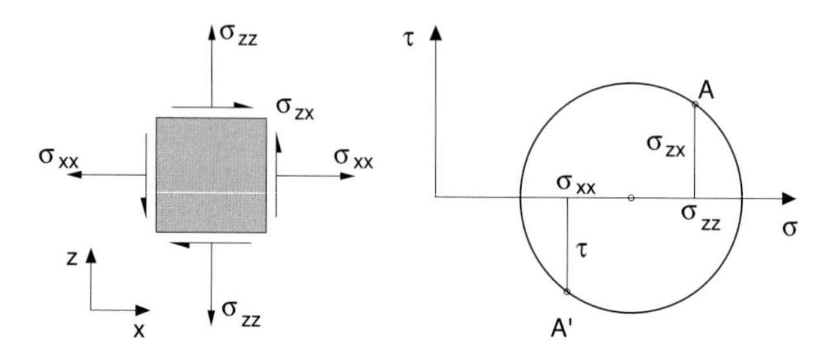

Figure 7.2 Stress components represented in a Mohr circle. Reproduced from [57], courtesy of Springer Nature.

Exercise 7.1.1 The tensile force in a rod is 60 N. Its cross-sectional area is 2 cm^2. The rod has the direction of the vector (1 1 2). Determine the stress tensor in the rod.

7.2 Mohr Circle

In cases where two components are sufficient to describe the stress state (e.g. in the triaxial test), the circle of Mohr offers a graphical representation that facilitates understanding and operating with the stress tensor (Fig. 7.2). We use a diagram where abscissa and ordinate show normal and shear stresses, respectively. The stress state is represented by a circle passing through the maximum and minimum principal stresses, σ_1 and σ_2, respectively. Its centre lies in the abscissa. Clearly, for a hydrostatic state $\sigma_2 = \sigma_1$, the stress circle shrinks to a point. The diameter $(\sigma_1 - \sigma_2)$, or the radius $(\sigma_1 - \sigma_2)/2$, of the Mohr circle indicate the stress deviator.

A particular point P on the circle of Mohr acts as pole in the following sense: A line l through P cuts the circle at a point the abscissa and ordinate of which indicate the normal and shear stresses, respectively, that act upon a plane parallel to l (Figs. 7.3 and 7.4).

7.3 Principal Stress Space

For rectilinear extensions (i.e. deformations with unchanged principal directions of strain), the stress is sufficiently described by the three principal stresses $\sigma_1, \sigma_2, \sigma_3$ that

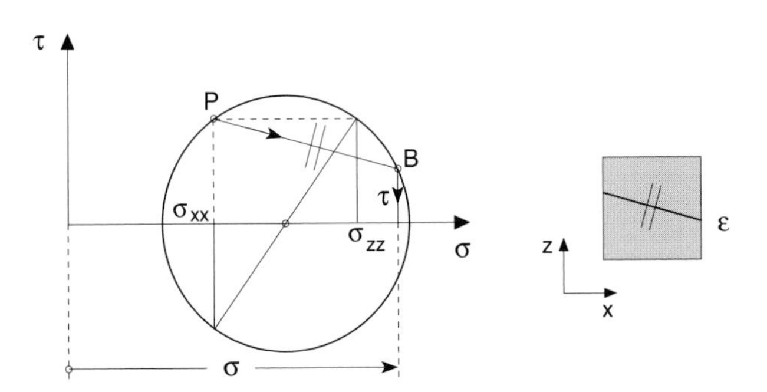

Figure 7.3 A line through the pole P having the direction of a plane ε cuts the Mohr circle at point B with abscissa σ and ordinate τ, the normal and tangential components of the stress vector acting upon that plane. Reproduced from [57], courtesy of Springer Nature.

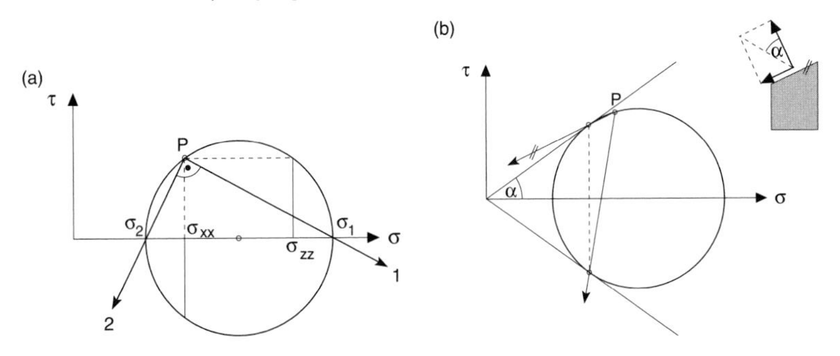

Figure 7.4 (a) The two lines through the pole P and the abscissas σ_1 and σ_2 have the directions of the principal stresses. They are perpendicular to each other. (b) The two lines through the pole P show the directions of maximum stress obliquity. Reproduced from [57], courtesy of Springer Nature.

can be represented as a point in a three-dimensional space. As many relations in soil mechanics are represented geometrically, it is worthwhile getting acquainted with this space. The line $\sigma_1 = \sigma_2 = \sigma_3$, the principal diagonal, is the locus of hydrostatic stresses. A plane π that passes through a stress point $\mathbf{T}$ and is perpendicular to the principal diagonal is called a deviatoric plane. Its distance from the origin is $|\mathrm{tr}\mathbf{T}|/\sqrt{3}$. The intersection of the principal diagonal with π has the distance $\frac{1}{3}\sqrt{\mathrm{tr}\mathbf{T}^{*2}}$ from $\mathbf{T}$. The angles α and ψ (Fig. 7.5) can be obtained as follows:

$$\cos(3\alpha) = \sqrt{6}\frac{\mathrm{tr}\,(\mathbf{T}^{*3})}{(\mathrm{tr}\,\mathbf{T}^{*2})^{3/2}}; \quad \cot\psi = \frac{\mathrm{tr}\,\mathbf{T}}{\sqrt{3\mathrm{tr}\,(\mathbf{T}^{*2})}}. \tag{7.1}$$

For some applications it is convenient to define Cartesian coordinates ξ and η in a deviatoric plane:

$$\xi = \frac{1}{\sqrt{2}}(\sigma_2 - \sigma_1); \qquad \eta = \frac{1}{\sqrt{6}}(2\sigma_3 - \sigma_1 - \sigma_2). \tag{7.2}$$

7.4 Stress Tensor in Cylindrical Coordinates

Cylindrical coordinates are advantageous for problems with cylindrical symmetry. The transformation rules from Cartesian to cylindrical coordinates are:

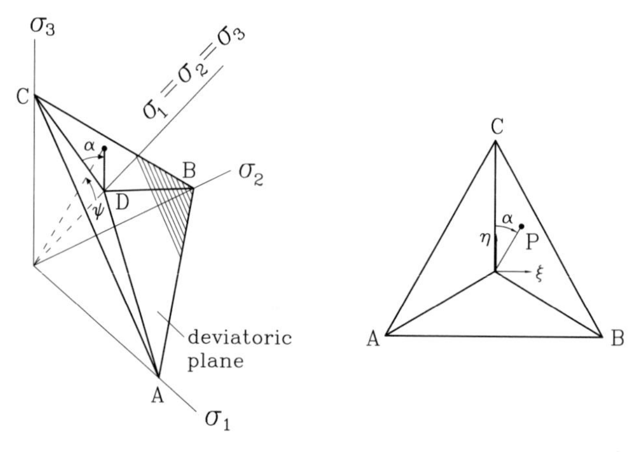

Figure 7.5 Principal stress space and deviatoric plane. The location of a point (= stress state) in the principal stress space can be given by the angles α, ψ and the distance of the deviatoric plane from the origin. Reproduced from [57], courtesy of Springer Nature.

$$
\begin{aligned}
\sigma_{xx} &= \frac{\sigma_{rr} + \sigma_{\theta\theta}}{2} + \frac{\sigma_{rr} - \sigma_{\theta\theta}}{2} \cos 2\theta - \sigma_{r\theta} \sin 2\theta \\
\sigma_{yy} &= \frac{\sigma_{rr} + \sigma_{\theta\theta}}{2} - \frac{\sigma_{rr} - \sigma_{\theta\theta}}{2} \cos 2\theta + \sigma_{r\theta} \sin 2\theta \\
\sigma_{xy} &= \frac{\sigma_{rr} - \sigma_{\theta\theta}}{2} \sin 2\theta + \sigma_{r\theta} \cos 2\theta
\end{aligned}
$$

and

$$
\begin{aligned}
\sigma_{rr} &= \frac{\sigma_{xx} + \sigma_{yy}}{2} + \frac{\sigma_{xx} - \sigma_{yy}}{2} \cos 2\theta + \sigma_{xy} \sin 2\theta \\
\sigma_{\theta\theta} &= \frac{\sigma_{xx} + \sigma_{yy}}{2} - \frac{\sigma_{xx} - \sigma_{yy}}{2} \cos 2\theta - \sigma_{xy} \sin 2\theta \\
\sigma_{r\theta} &= -\frac{\sigma_{xx} - \sigma_{yy}}{2} \sin 2\theta + \sigma_{xy} \cos 2\theta .
\end{aligned}
$$

7.5 Invariants and Eigenvalues

This section refers to every tensor, not only to the stress tensor. The numerical values of the components of a tensor depend on the coordinate system. The invariants are numbers adjoint to a tensor that remain unchanged upon changes of the coordinate system. Considering the principal values T_1, T_2, T_3 of a tensor $\mathbf{T}$, the quantities $I_1 = \mathrm{tr}\mathbf{T} = T_1 + T_2 + T_3$, $I_2 = T_1T_2 + T_1T_3 + T_2T_3$ and $I_3 = \det \mathbf{T} = T_1T_2T_3$ are invariants. Also, the quantities $J_1 = \mathrm{tr}\mathbf{T}$, $J_2 = \mathrm{tr}\mathbf{T}^2$ and $J_3 = \mathrm{tr}\mathbf{T}^3$ are invariants. They are related via the equations $J_2 = I_1^2 - 2I_2$ and $J_3 = I_1^3 - 3I_1I_2 + 3I_3$. Clearly, any algebraic expression formed with these invariants is also an invariant.

We consider the case that the stress vector $\mathbf{t}$ upon a plane has the direction of the unit normal vector $\mathbf{n}$. The then valid equation $\mathbf{Tn} = \lambda\mathbf{n}$ leads to $(\mathbf{T} - \lambda\mathbf{1})\mathbf{n} = \mathbf{0}$, hence $\det(\mathbf{T} - \lambda\mathbf{1}) = 0$. This algebraic equation is equivalent to

$$
-\lambda^3 + I_1\lambda^2 - I_2\lambda + I_3 = 0 \tag{7.3}
$$

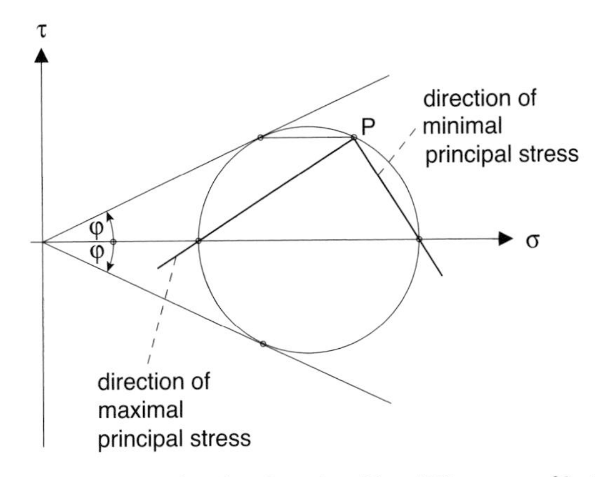

Figure 7.6 Principal stresses at the limit state in a shear box. Reproduced from [57], courtesy of Springer Nature.

and has the solutions $\lambda_1 = T_1, \lambda_2 = T_2, \lambda_3 = T_3$, where T_i are the so-called principal values of the tensor **T** (also called eigenvalues or principal stresses).

Regarding the above stated polynomial equation, the Cayley–Hamilton theorem states that it is also fulfilled by **T**:

$$-\mathbf{T}^3 + I_1\mathbf{T}^2 - I_2\mathbf{T} + I_3\mathbf{1} = \mathbf{0}. \tag{7.4}$$

Consequently, the third power of **T** (and hence also all higher powers) can be expressed as linear combinations of $\mathbf{1}, \mathbf{T}$ and $\mathbf{T}^2$. This results in the following representation theorem: A tensor-valued function $\mathbf{f}(\mathbf{T})$ of a symmetric tensorial argument **T** can be represented as

$$\mathbf{f}(\mathbf{T}) = \phi_0\mathbf{1} + \phi_1\mathbf{T} + \phi_2\mathbf{T}^2, \tag{7.5}$$

where ϕ_1, ϕ_2 and ϕ_3 depend on invariants of **T**. Representation theorems are important for the formulation of constitutive equations, see Section 12.2.1.

7.6 Example: Stress in a Shear Box

What is the direction of principal stresses within the shear band between the upper and lower halves of a shear box? In the horizontal shear band, we have $\tau/\sigma = \tan\phi$. From the point where the circle touches the line $\tau = \sigma \cdot \tan\phi$ we draw a horizontal line that cuts the circle at the pole P (Fig. 7.6). From there we draw the lines to the points where the circle cuts the abscissa. These are the directions of the principal stresses.

8 Conservation Laws (Balance Equations)

8.1 Integrals of Motion

Integrals of motion are quantities such as energy and momentum (impulse) that remain unchanged during mechanical processes, and this is expressed by the so-called conservation laws or balance equations. Conservation of energy results from the invariance of physical laws with respect to time, and conservation of momentum results from invariance with respect to space (the theorem of Noether). Conservation of mass is a special case of energy conservation for non-relativistic cases. In view of the dissipative character of soil deformation, the conservation of energy is not considered here, because the amount of energy transformed to heat is usually not quantified in soil mechanics. So, we restrict our attention to the conservation of mass and momentum.

Conservation laws can be formulated as

- field equations
- jump relations
- integral equations.

Although the terms 'conservation equation' and 'balance equation' are often used as synonyms, the following distinction can be made. The general statement of a balance equation is that the production of a quantity equals the sum of its storage and outflow. A conservation equation refers to the special case that the production term vanishes [106].

8.2 Conservation Laws as Field Equations

Conservation relations can be expressed by field equations, i.e. differential equations for the field variables [116]. This presupposes that the derivatives of the involved variables exist and are continuous throughout the field.

8.2.1 Mass Conservation as a Field Equation

With ρ being the density, mass conservation reads:

$$\dot{\rho} + \rho \operatorname{div} \mathbf{v} = 0. \tag{8.1}$$

With $\dot{\rho} = \frac{\partial \rho}{\partial t} + \frac{\partial \rho}{\partial \mathbf{x}} \mathbf{v}$ we obtain

$$\frac{\partial \rho}{\partial t} + \operatorname{div}(\rho \mathbf{v}) = 0. \tag{8.2}$$

The mass conservation for incompressible fluids is the so-called continuity equation $\operatorname{div} \mathbf{v} = 0$.

8.2.2 Conservation of Momentum as a Field Equation

The impulse (or momentum) balance equation is vectorial, it corresponds to three scalar equations. It can be written in the following forms:

$$\frac{\partial(\rho \mathbf{v})}{\partial t} + \operatorname{div}\left(\rho \mathbf{v} \otimes \mathbf{v} - \mathbf{T}\right) = \rho \mathbf{b} \tag{8.3}$$

$$\frac{\partial(\rho v_i)}{\partial t} + \frac{\partial}{\partial x_j}(\rho v_i v_j - T_{ij}) = \rho b_i \tag{8.4}$$

$$\rho \frac{\partial v_i}{\partial t} + \frac{\partial \rho}{\partial t} v_i + \frac{\partial \rho}{\partial x_j} v_i v_j + \rho \frac{\partial v_i}{\partial x_j} v_j + \rho v_i \frac{\partial v_j}{\partial x_j} - \frac{\partial}{\partial x_j} T_{ij} = \rho b_i \tag{8.5}$$

$$v_i \underbrace{\left(\frac{\partial \rho}{\partial t} + \frac{\partial \rho}{\partial x_j} v_j + \rho \frac{\partial v_j}{\partial x_j}\right)}_{=0,\ \text{due to mass balance}} + \rho \underbrace{\left(\frac{\partial v_i}{\partial t} + \frac{\partial v_i}{\partial x_j} v_j\right)}_{\dot{v}_i} - \frac{\partial}{\partial x_j} T_{ij} = \rho b_i. \tag{8.6}$$

Taking into account mass balance, we thus have:

$$T_{kl,l} + \rho b_k = \rho \ddot{x}_k \tag{8.7}$$

and in symbolic notation:

$$\nabla \cdot \mathbf{T} + \rho \mathbf{b} = \rho \ddot{\mathbf{x}}, \tag{8.8}$$

where $\rho \mathbf{b}$ is the force per unit of volume ('volume force') applied by an external field, $\ddot{\mathbf{x}}$ is the acceleration of a particle $\mathbf{X}$ and has also the meaning of a force per unit of mass ('mass force'). Equation 8.7 or Equation 8.8 is called *the first Cauchy equation of motion*. For the so-called quasistatic case, $\ddot{\mathbf{x}} = \mathbf{0}$, it is called the *equilibrium equation*.

Remark Directional blasting. From Equation 8.8 follows an interesting application to the blasting of soil in a desired direction, e.g. to quickly create a dam for protection against an imminent flood, according to an idea of Lavrentief [63]. As is known, an explosive is an intimate mixture of a combustible substance and an oxygen supplier. On ignition, combustion (i.e. oxidation) occurs abruptly and transforms the explosive into a gas mixture which is initially under a very high pressure p. For gases and for $\mathbf{b} = \mathbf{0}$, Equation 8.8 reads $-\nabla p = \rho \dot{\mathbf{v}}$. With the 'pressure impulse' $P = \int_0^t p dt$ it follows: $\mathbf{v} = -\frac{1}{\rho} \nabla P$ so that $\mathbf{v} = $ const can be achieved if P is a linear function of position $\mathbf{x}$, i.e. $P = ax+by+cz+d$. Assuming that the pressure impulse is proportional to the mass of explosive, we only need to distribute it linearly along the boundary of the soil to be expelled, see Fig. 8.1. The underlying idea is that the distribution of the velocity potential P within the considered region of soil is as prescribed on its boundary.

Figure 8.1 Distribution of explosive for directional blasting.

8.2.3 Conservation of Moment of Momentum

This conservation law implies the symmetry of the stress tensor

$$\mathbf{T} = \mathbf{T}^T, \tag{8.9}$$

if couple stresses are not taken into account. Couple stresses are relevant for so-called higher continua, such as Cosserat continua.

8.2.4 Couple Stresses

Boltzmann pointed to the fact that the symmetry of the stress tensor is nothing but an axiom. Therefore, continua with non-symmetric stress tensors (so-called non-Boltzmann continua) are conceivable, though their practical value is limited. Let $\mathrm{d}m_j$ be a couple acting upon a surface element. $\mathrm{d}m_j$ is connected with the surface element $\mathrm{d}F_i$ by means of the *couple stress tensor* μ_{ij}:

$$\mathrm{d}m_j = \mu_{ij}\mathrm{d}F_i. \tag{8.10}$$

Consideration of equilibrium of moments yields

$$\partial_i\mu_{ij} + 2\sigma_j = -r_j.$$

Herein, σ_j is the vector corresponding to the antisymmetric part of the force–stress tensor: $\sigma_k = \frac{1}{2}e_{ijk}\sigma_{ij}$, $\sigma_{[ij]} = e_{ijk}\sigma_k$. σ_k vanishes only in the absence of couple stresses and couple density r_j on the boundary. At the boundary we have to assume forces (tractions) and moments distributed over the surface,

$$q_j = \sigma_{ij}n_i\ , \quad r_j = \mu_{ij}n_i\ ,$$

where n_i is the unit normal vector. Couple stress can also be considered as a tensor of third degree: μ_{nmj}. If we introduce Equation 4.15 into Equation 8.10, we obtain

$$\mathrm{d}m_j = \mu_{ij}\,\frac{1}{2}e_{nmi}\,\mathrm{d}F_{nm} = \mu_{nmj}\,\mathrm{d}F_{nm}.$$

The moment density r_i can be also considered as tensor of the second degree: $r_{nm} = e_{nmi}r_i$.

Table 8.1. Correspondence of Cosserat and beam theories

Quantity	Cosserat theory	beam theory
Rotation	ω_{ij}^{inc}	y'
Curvature	$\kappa_{k[ij]} = \partial\omega_{ij}^{inc}/\partial x_k, \quad \kappa_{ij} = \frac{1}{2}e_{lkj}\ \kappa_{i[lk]}$	y''
Couple stress	μ_{ij}	M
Antisymmetric stress	$\sigma_{[ij]} = e_{ijk}\sigma_k; \quad \sigma_k = \frac{1}{2}e_{ijk}\ \sigma_{ij}$	Q
Work	$dA = \sigma_{[ij]}d\omega_{ij}^{inc} + \mu_{ij}d\kappa_{ij}$	$dA = Qd(y') + Md(y'')$
Equilibrium	$\partial_i\mu_{ij} + 2\sigma_j = 0$	$\frac{dM}{dx} + Q = 0$

The energy-conjugate variables to μ_{ij} and σ_i are the curvature tensor κ_{ij} and the incompatible (or Cosserat) rotation ω_i^{inc}. Thus, the increment of deformation work reads

$$dA = \sigma_{ij}d\varepsilon_{ij} + \sigma_i d\omega_i^{inc} + \mu_{ij}d\kappa_{ij}.$$

The evolutions of σ_i und μ_{ij} with the kinematic variables $\varepsilon_{ij}, \omega_i^{inc}, \kappa_{ij}$ must be given by appropriate constitutive equations in addition to the one describing the evolution of stress σ_{ij}.

Interestingly, the linearised beam can be considered as a one-dimensional Cosserat medium. With the deflection $y(x)$, the bending moment M and the transversal force Q we have the analogy shown in Table 8.1 (in the absence of distributed bending moments along the beam). Here the beam deformation is considered as represented only by y and y' or ω_{ij}^{inc} and κ_{ij}, hence the part $\sigma_{ij}d\varepsilon_{ij}$ is omitted. Illustrative examples of oriented media are bird flocks and schools of fish if one considers them as continua.

Despite many publications, the utility of the Cosserat theory in soil mechanics remains limited, as couple stresses can hardly be measured. So, constitutive laws can hardly be designed, let alone calibrated. The same holds for the related boundary conditions.

8.3 Weak Solution of the Equilibrium Equation

By neglecting volume forces, the differential equation of equilibrium reads $\nabla \cdot \mathbf{T} = \mathbf{0}$ or $\sigma_{ij,j} = 0$. In the quasi-static case, it must be fulfilled *locally*, i.e. at every point of the considered body. A so-called weak solution of this differential equation is obtained when we merely pose a *global* requirement, namely that the following integral vanishes:

$$\int_V \sigma_{ij,j}\bar{v}_i\ dV = 0, \tag{8.11}$$

where $\bar{v}_i$ is a so-called test function and can also be considered as a *virtual velocity*. It need not be the real velocity; it only needs to fulfil the kinematical boundary conditions and internal kinematical constraints (if posed). Noting that

$$\text{div}(\bar{v}_i\sigma_{ij}) = \partial_j(\bar{v}_i\sigma_{ij}) = \bar{v}_{i,j}\sigma_{ij} + \bar{v}_i\sigma_{ij,j} \tag{8.12}$$

we observe that $\sigma_{ij,j}$ vanishes because of equilibrium, and with $\bar{v}_{i,j}\sigma_{ij} = \dot{\bar{\varepsilon}}_{ij}\sigma_{ij}$, we obtain (8.11) as:

$$\int_V \dot{\bar{\varepsilon}}_{ij}\sigma_{ij}\mathrm{d}V = 0. \tag{8.13}$$

This integral represents the virtual power of the stress σ_{ij} integrated over the volume of the body. Applying the divergence theorem, we obtain that this equals the power of the tractions acting upon the surface of the body. This equality is known as the principle of virtual power, or principle of virtual work (PVW). The latter is obtained if we replace the virtual velocity $\bar{\mathbf{v}}$ by the virtual displacement $\bar{\mathbf{u}}$ and also $\dot{\bar{\varepsilon}}$ by $\bar{\varepsilon}$.

$$\underbrace{\int_V \dot{\bar{\varepsilon}}_{ij}\sigma_{ij}\mathrm{d}V}_{\textit{internal virtual power}} = \underbrace{\int_S (\dot{\bar{v}}_i\sigma_{ij})n_j\mathrm{d}S}_{\textit{external virtual power}} . \tag{8.14}$$

This principle is equivalent to the equilibrium equation. In the method of finite elements, it is used to assign a nodal force resulting from the deformation of the adjacent finite elements. This so-called out-of-balance force must vanish in the case of static equilibrium.

8.4 Jump Relations

Field equations presuppose continuous fields and, hence, cannot be applied at fields with discontinuities (see also Section 5.4). Across discontinuity surfaces with unit normal vector $\mathbf{n}$, which possibly move with normal velocity $\mathbf{u}$, the conservation laws are expressed as jump relations [6] (Table 8.2):

- conservation of mass: $[\rho\,(\mathbf{u} - \dot{\mathbf{x}})] \cdot \mathbf{n} = 0$
- conservation of impulse: $[\rho\dot{\mathbf{x}}\,(\mathbf{u} - \dot{\mathbf{x}}) \cdot \mathbf{n} + \mathbf{Tn}] = \mathbf{0}$
- conservation of energy: $\left[\rho\left(e + \frac{\mathbf{v}^2}{2}\right)(\mathbf{u} - \mathbf{v}) + (\mathbf{Tv} - \mathbf{q})\right] \cdot \mathbf{n} = 0$. e is the specific internal energy, i.e. the internal energy per unit volume. As is known, internal energy is that part of the energy of a body which does not depend on the motion of the observer. $\mathbf{q}$ is the heat flux.

  This equation is relevant, e.g. for soil freezing, which is a standard method of temporarily stabilising soil. A cool fluid is circulated within pipes driven into the soil. As a result, the soil surrounding the pipes freezes, and the outer boundary of the frozen soil cylinder expands with velocity $\mathbf{u}$. This boundary is a moving discontinuity surface and e is the latent heat of water (334.5 kJ/kg).

Exercise 8.4.1 Sedimentation front. Silt particles float in a lake so that the water has a specific weight of 11 kN/m^3. They sink with a velocity of 1 cm per 100 years. The bottom of the lake is silt with specific weight of 20 kN/m^3. Determine the velocity with which the bottom of the lake rises.

Table 8.2. Conservation equations as field equations and jump relations

	Mass	Impulse
Field equation	$\frac{\partial\rho}{\partial t} + \mathrm{div}(\rho\dot{\mathbf{x}}) = 0$	$\frac{\partial(\rho\mathbf{v})}{\partial t} + \mathrm{div}\left(\rho\mathbf{v}\otimes\mathbf{v} - \mathbf{T}\right) = \rho\mathbf{b}$
Jump relation	$[\rho\,(\mathbf{u} - \dot{\mathbf{x}})] \cdot \mathbf{n} = 0$	$[\rho \cdot \dot{\mathbf{x}}\,(\mathbf{u} - \dot{\mathbf{x}}) \cdot \mathbf{n} + \mathbf{Tn}] = \mathbf{0}$

8.5 Integral Representations of Conservation Equations

The conservation of a quantity $\Phi(\mathbf{x}, t)$ can also be expressed as the balance of Φ integrated over a volume V. To do this, we need the total time derivative of $\int_V \Phi dV$, which is given by the Reynolds transport theorem. Depending on whether V is a material volume or a spatially fixed (or moving) control volume, there are various forms of the Reynolds transport theorem. For V being a material volume, the Reynolds transport theorem reads:

$$\frac{\mathrm{d}}{\mathrm{d}t}\int_V \Phi\, dV = \int_V \left(\frac{\partial \Phi}{\partial t} + \operatorname{div}(\Phi \mathbf{v})\right) \mathrm{d}V \tag{8.15}$$

or, using the theorem of Gauss (also called divergence theorem) and the outwards oriented normal vector $\mathbf{n}$ on the boundary:

$$\frac{\mathrm{d}}{\mathrm{d}t}\int_V \Phi\, dV = \int_V \frac{\partial \Phi}{\partial t}\mathrm{d}V + \int_{\partial V} \Phi \mathbf{v}\cdot\mathbf{n}\,\mathrm{d}A. \tag{8.16}$$

With $\Phi = \rho$ and mass $m = \int_V \rho\ \mathrm{d}V$ we obtain from Equation 8.5 the integral representation of mass conservation:

$$\begin{aligned}\frac{\mathrm{d}m}{\mathrm{d}t} &= \int_V \left(\rho' + \operatorname{div}(\rho\,\mathbf{v})\right)\,\mathrm{d}V = \int_V \left(\rho' + \mathbf{v}\cdot\nabla\rho + \rho\operatorname{div}\mathbf{v}\right)\,\mathrm{d}V \\ &= \int_V (\dot\rho + \rho\,\operatorname{div}\mathbf{v})\,\mathrm{d}V = 0\end{aligned} \tag{8.17}$$

with $\rho'(\mathbf{x}, t) = \dfrac{\partial\rho(\mathbf{x}, t)}{\partial t}$ and $\dot\rho$ being the material (or 'total') time derivative of ρ. From Equations 8.16 and 8.17 follows:

$$\int_V \rho' \mathrm{d}V = -\int_{\partial V} \rho\mathbf{v}\cdot\mathbf{n}\,\mathrm{d}A.$$

The integral representation of conservation of momentum reads:

$$\int_{\partial V} \mathbf{T}\mathbf{n}\,\mathrm{d}A + \int_V \mathbf{b}\,\mathrm{d}V = \int_V \dot{\mathbf{v}}\rho\,\mathrm{d}V, \tag{8.18}$$

where $\mathbf{b}$ is the mass force (e.g. gravity acceleration). With the Gauss theorem we obtain

$$\int_V (\operatorname{div}\mathbf{T} + \mathbf{b} - \rho\dot{\mathbf{v}})\,\mathrm{d}V = 0. \tag{8.19}$$

8.6 Stress and Intergranular Forces

The so-called Discrete Element Method considers soil as an assembly of individual grains and their interactions. The intergranular forces can be related to a mean stress as follows: We multiply the equilibrium equation $\frac{\partial \sigma_{il}}{\partial x_l} = 0$ with x_k and integrate over a volume. Applying the product rule yields:

$$\int \frac{\partial \sigma_{il}}{\partial x_l} x_k \mathrm{d}V = \int \frac{\partial(\sigma_{il} x_k)}{\partial x_l} \mathrm{d}V - \int \sigma_{il} \frac{\partial x_k}{\partial x_l} \mathrm{d}V = 0.$$

With the divergence theorem we transform the first volume integral into a surface integral and simplify the second integral using $\partial x_k / \partial x_l = \delta_{kl}$. We thus obtain:

$$\int \sigma_{il} x_k \mathrm{d}S_l - \int \sigma_{ik} \mathrm{d}V = 0.$$

Considering discrete forces, the surface integral is equivalent to $\sum(\mathbf{P}_i \mathbf{x}_i)$. Herein, $\mathbf{P}_i$ is a force acting upon the surface of the considered body and $\mathbf{x}_i$ is the position vector to the point of its application. With $\bar{\sigma}_{ik} = \frac{1}{V} \int \sigma_{ik} dV$ we finally obtain the mean stress $\bar{\sigma}_{ik}$ expressed via the intergranular forces $\mathbf{P}_i$:

$$\bar{\sigma}_{ik} = \frac{1}{V} \sum_i (\mathbf{P}_i \mathbf{x}_i).$$

In a similar way we consider the displacement gradient $\nabla \mathbf{u}$ and obtain:

$$\bar{\varepsilon}_{ij} = \overline{\frac{1}{2}\left(\frac{\partial u_i}{\partial x_j} + \frac{\partial u_j}{\partial x_i}\right)} = \frac{1}{V} \sum u_i x_j.$$

8.7 Stress Fields

For quasi-static cases, stress fields are tensor-valued functions of position $\mathbf{x}$ that must fulfil the equations of equilibrium. Some examples of stress fields are given here. A convenient representation of stress fields in cases of plane deformation is given by plots of the *principal stress trajectories* (Fig. 8.2). These are lines in the directions of the principal stresses. Plane deformation means that there is no change perpendicular to a plane and, consequently, no shear stresses acting upon that plane.

8.7.1 Examples

Geostatic Stress Field

A geostatic stress field prevails within a homogeneous halfspace with horizontal surface and constant specific weight γ. We assume that the soil stratum is generated by sedimentation which implies a gradual increase of the weight. The related deformation is a one-dimensional compression and corresponds to the deformation in an oedometer. Consequently, the ratio of the horizontal to the vertical stresses is for virgin loading equal to K_0. For un- and reloading this ratio is increased, as shown in Fig. 16.11. Unloading of a soil stratum occurs e.g. when an overlaying glacier melts. With the z-coordinate pointing downwards the principal stresses are:

$$\sigma_z = \gamma z; \qquad \sigma_x = K_0 \sigma_z; \quad K_0 = \text{const} \tag{8.20}$$

The vertical and horizontal stresses are principal stresses because the equilibrium condition in the x direction implies: $\frac{\partial \sigma_{xz}}{\partial z} + \frac{\partial \sigma_{xx}}{\partial x} = 0$. Because of the infinite extension

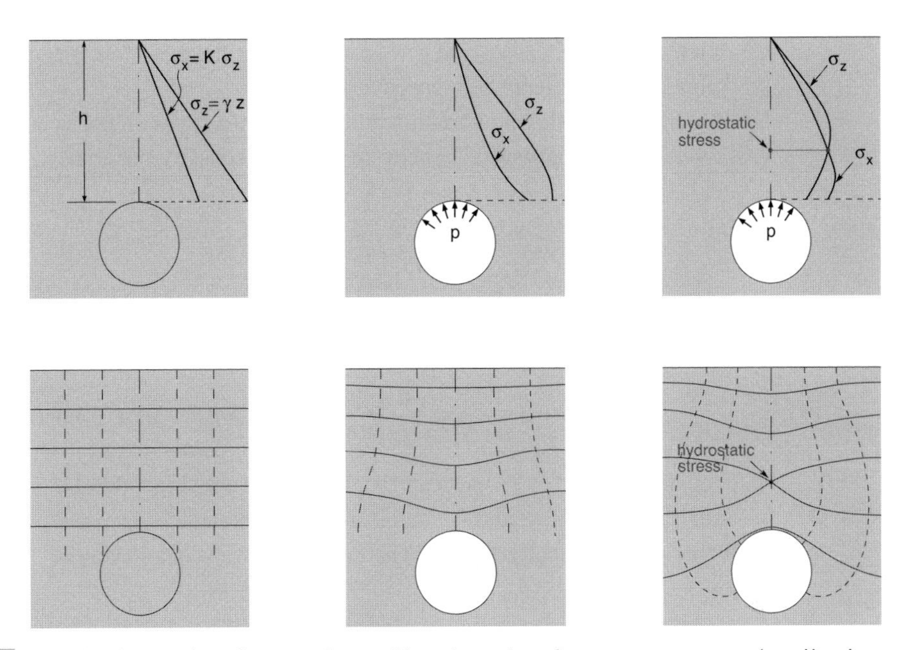

Figure 8.2 Stress distributions above the crown of a tunnel for various values of support pressure $p = \sigma_z(z = h)$ and corresponding stress trajectories. At the hydrostatic point (right), the curvature $1/r$ of the stress trajectory vanishes, $r = \infty$. Reproduced from [57], courtesy of Springer Nature.

in the horizontal direction x, the derivative $\partial/\partial x$ vanishes, thus it follows: $\sigma_{xz} = \text{const}$ If $\sigma_{xz} = 0$ applies on the surface $z = 0$, then σ_{xz} vanishes everywhere.

Stress Field in Shaled Rock

We consider a halfspace with horizontal surface, which consists of a slate with inclined orientation. The slate planes (unit normal vector $\mathbf{n}$) can – by assumption – not transfer shear stresses. The prevailing stress state can be given as follows: The stress vector must be normal to the slate surface: $\mathbf{Tn} = \lambda\mathbf{n}$, as with the eigenvalue problem. It follows that $\mathbf{n}$ must be an eigendirection of $\mathbf{T}$.

8.7.2 Equilibrium Equations in Other Systems of Coordinates

The equilibrium equations can be written in any arbitrary system of coordinates. For some problems, particular systems of coordinates are advantageous:

Cylindrical coordinates: The equilibrium equations in cylindrical coordinates read:

$$\frac{\partial \sigma_{rr}}{\partial r} + \frac{1}{r} \cdot \frac{\partial \sigma_{\theta r}}{\partial \theta} + \frac{\partial \sigma_{zr}}{\partial z} + \frac{1}{r} \cdot (\sigma_{rr} - \sigma_{\theta\theta}) + \varrho b_r = \varrho a_r$$

$$\frac{\partial \sigma_{r\theta}}{\partial r} + \frac{1}{r} \cdot \frac{\partial \sigma_{\theta\theta}}{\partial \theta} + \frac{\partial \sigma_{z\theta}}{\partial z} + \frac{1}{r} \cdot (\sigma_{r\theta} + \sigma_{\theta r}) + \varrho b_\theta = \varrho a_\theta$$

$$\frac{\partial \sigma_{rz}}{\partial r} + \frac{1}{r} \cdot \frac{\partial \sigma_{\theta z}}{\partial \theta} + \frac{\partial \sigma_{zz}}{\partial z} + \frac{1}{r} \cdot \sigma_{rz} + \varrho b_z = \varrho a_z,$$

where $\boldsymbol{b} = \{b_r, b_\theta, b_z\}$ and $\boldsymbol{a} = \{a_r, a_\theta, a_z\}$ are the mass forces (i.e. force per unit mass) and acceleration, respectively.

Slip lines: The slip lines are lines along which the stress obliquity has its maximum value. The equations of Hencky are valid for cohesive materials:

$$\frac{\partial \sigma_m}{\partial s_1} - 2c\frac{\partial \alpha}{\partial s_1} = 0, \quad \frac{\partial \sigma_m}{\partial s_2} + 2c\frac{\partial \alpha}{\partial s_2} = 0. \tag{8.21}$$

The equations of Kötter and Massau are valid for frictional materials:

$$\frac{\partial \sigma_m}{\partial s_1} - 2\sigma_m \tan\varphi \frac{\partial \alpha}{\partial s_1} = \frac{1}{\cos\varphi}[\gamma_x \sin(\alpha+\mu) - \gamma_z \cos(\alpha+\mu)]$$

$$\frac{\partial \sigma_m}{\partial s_1} + 2\sigma_m \tan\varphi \frac{\partial \alpha}{\partial s_2} = \frac{1}{\cos\varphi}[-\gamma_x \sin(\alpha-\mu) + \gamma_z \cos(\alpha-\mu)],$$

with $\sigma_m := \frac{1}{2}(\sigma_x + \sigma_z)$; $\alpha :=$ angle between the direction of σ_1 and the x-axis; $\mu := 45^\circ - \varphi/2$. s_1 and s_2 are the arc lengths along the slip lines.

Principal stress trajectories: The Lamé–Maxwell equations read:

$$\frac{\partial \sigma_1}{\partial s_1} + \frac{\sigma_1 - \sigma_2}{r_2} + X_1 = 0\,; \quad \frac{\partial \sigma_2}{\partial s_2} + \frac{\sigma_1 - \sigma_2}{r_1} + X_2 = 0, \tag{8.22}$$

where σ_1 and σ_2 are the principal stresses, s_1 and s_2 are the arc lengths on the corresponding trajectories and r_1 and r_2 are the radii of curvature of these trajectories (positive for curves turning left). X_1 and X_2 are the volume forces in the directions of the trajectories s_1 and s_2.

8.7.3 Example: Stress Field around a Tunnel, Arching

Near the apex (crown) of a tunnel, axial symmetry prevails (because of $\partial/\partial\vartheta = 0$). We can therefore locally use cylindrical coordinates.

The stress components σ_r, σ_θ, σ_z are principal stresses. The equation of equilibrium in the r-direction reads:

$$\frac{\partial \sigma_r}{\partial r} + \frac{\sigma_r - \sigma_\theta}{r} + \varrho \boldsymbol{g} \cdot \boldsymbol{e}_r = 0. \tag{8.23}$$

Herein, ϱ is the density, $\varrho \boldsymbol{g}$ is the unit weight and $\mathbf{e}_r$ is the unit vector in the r-direction. The second term in Equation 8.23 describes arching.

If r increases in the vertical direction z, then Equation 8.23 reads in Cartesian coordinates:

$$\frac{\mathrm{d}\sigma_z}{\mathrm{d}z} = \gamma - \frac{\sigma_x - \sigma_z}{r}. \tag{8.24}$$

Herein, the term $(\sigma_x - \sigma_z)/r$ is responsible for the fact that σ_z does not increase linearly with depth (i.e. $\sigma_z = \gamma z$). In the case of arching, i.e. for $(\sigma_x - \sigma_z)/r > 0$, σ_z increases underproportionally with the depth z. This term, and thus arching, exists only for $\sigma_r \neq \sigma_\theta$. This means that arching is due to the ability of a material to sustain deviatoric stress, i.e. shear stress. No arching is possible in fluids. This is why soil/rock often 'forgives' shortages of support, whereas (ground)water is merciless.

The arching term in Equation 8.23 can be explained as follows: Consider the volume element shown in Fig. 8.3. The resultant A of the radial stresses reads

$$A = (\sigma_r + \mathrm{d}\sigma_r)\cdot(r + \mathrm{d}r)\mathrm{d}\theta - \sigma_r r\mathrm{d}\theta \approx \sigma_r \mathrm{d}r\mathrm{d}\theta + \mathrm{d}\sigma_r r\mathrm{d}\theta \tag{8.25}$$

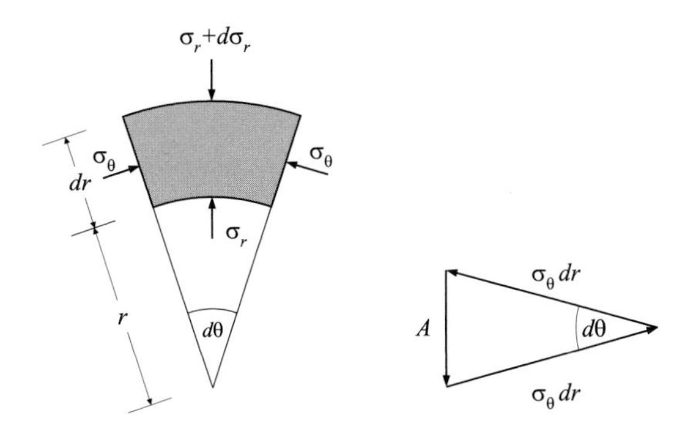

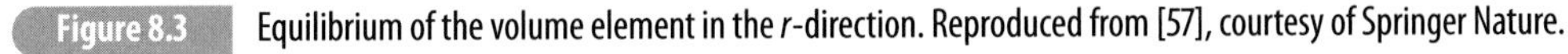
Figure 8.3 Equilibrium of the volume element in the r-direction. Reproduced from [57], courtesy of Springer Nature.

and should counterbalance the resultant of the tangential (or hoop) stresses σ_θ. The vectorial sum of the forces shown in Fig. 8.3 yields

$$\frac{A}{\sigma_\theta \mathrm{d}r} = \mathrm{d}\theta. \tag{8.26}$$

It then follows (from Equation 8.25, for $\mathbf{g} \cdot \mathbf{e}_r = 0$):

$$\frac{\mathrm{d}\sigma_r}{\mathrm{d}r} + \frac{\sigma_r - \sigma_\theta}{r} = 0. \tag{8.27}$$

With reference to the arching term $\frac{\sigma_\theta - \sigma_r}{r}$, attention should be paid to r. At the tunnel crown (Fig. 8.2), r is often set equal to the curvature radius of the crown. If we consider the distributions of σ_z and σ_x above the crown at various support pressures p, we notice that for $K = \sigma_x/\sigma_z < 1$ the horizontal stress trajectory has the opposite curvature than the tunnel crown.

9 Internal Friction and Shear Strength

9.1 Meaning of Strength

Tensile and shear stresses cannot be increased beyond some limit called strength. For soils, tensile strength is very small or 0, so only shear strength counts. It equals the maximum applicable shear stress and is attributed mainly to friction, which is originally a concept for rigid bodies in contact with each other. This concept goes back to Leonardo da Vinci, Amontons and Coulomb and states that bodies do not slide relative to each other if the inclination φ_m of the interaction force is smaller or equal to the value φ, the friction angle. In soil mechanics, collapse is often manifested as the formation of soil blocks that slide relative to each other without being deformed (i.e. as rigid bodies). In this way, the concept of friction can also be applied to soil mechanics.

Strength matters whenever we want to prove that collapse is not imminent for a given situation, or when we want to analyse the occurrence of a collapse.

9.2 Dry Friction in Continuum Mechanics

To transfer the concept of friction to continuum mechanics, we consider the normal stress σ and the shear stress τ acting upon an infinitesimal cut surface. The force inclination corresponds now to the *stress obliquity* τ/σ, which must not exceed the value $\tan\varphi$, where φ is the angle of internal friction (Fig. 9.1). Alternatively, we may write $\varphi_m \leq \varphi$ with the mobilised friction angle $\varphi_m := \arctan(\tau/\sigma)$. Using Mohr's stress circle we can easily express this condition in terms of the maximum and minimum principal stresses σ_1 and σ_3, respectively. For the case $\varphi_m = \varphi$, the Mohr circle with centre at $(\sigma_1 + \sigma_3)/2$ and diameter $\sigma_1 - \sigma_3$ is tangent to the line $\tau = \sigma \cdot \tan\varphi$ (Fig. 9.2), hence

$$\sin\varphi = \frac{\sigma_1 - \sigma_3}{\sigma_1 + \sigma_3}. \tag{9.1}$$

The intermediate principal stress σ_2 is not important, because if $\sigma_2 < \sigma_3$ or $\sigma_2 > \sigma_1$, then $\varphi_a = \arcsin\frac{\sigma_1-\sigma_2}{\sigma_1+\sigma_2}$ or $\varphi_b = \arcsin\frac{\sigma_2-\sigma_3}{\sigma_2+\sigma_3}$ would be larger than φ, which – by assumption – is not possible. Thus, if equation 9.1 is fulfilled, the following must apply: $\sigma_3 \leq \sigma_2 \leq \sigma_1$.

Equation 9.1 is a condition for failure manifested as the sliding of blocks relative to each other. Herein, failure denotes the increase of deformations or displacements at constant stresses or external loads. Of course, the condition (Equation 9.1) must

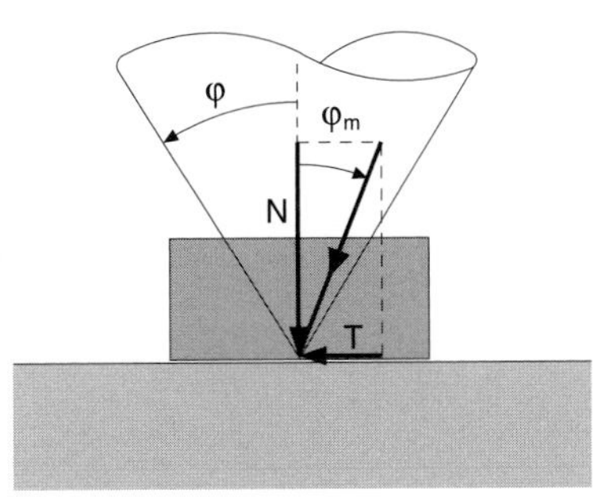

Figure 9.1 The rigid block will not slip if $\varphi_m < \varphi$. Reproduced from [57], courtesy of Springer Nature.

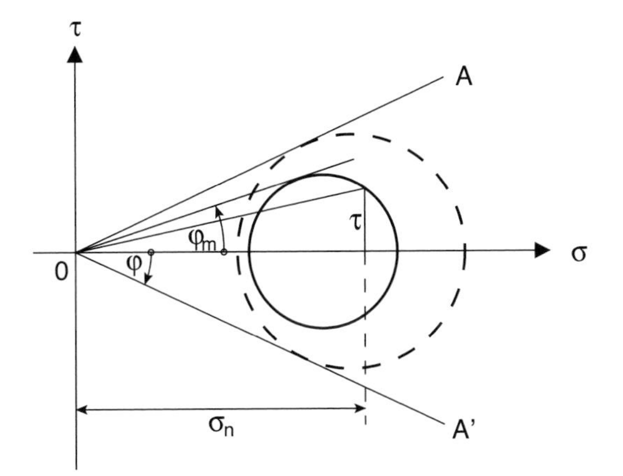

Figure 9.2 The circle represents a stress state with maximum stress obliquity φ_m. The dashed circle represents a limit stress state. Reproduced from [57], courtesy of Springer Nature.

apply not only at one point but at a collection of points such that a failure (or collapse) mechanism is formed that characterises a macroscopic failure. In most cases, collapse mechanisms consist of blocks that remain undeformed ('rigid') and slide relative to each other along slip surfaces. The latter are thin shear bands. An alternative (but less realistic) concept is that the failure condition is not localised within thin shear bands but is fulfilled within certain zones.

Stress states, which fulfil the equation $\tau = \sigma \cdot \tan\varphi$ or Equation 9.1, form a surface in the space of principal stresses, the so-called limit state (or limit condition) according to Mohr–Coulomb. Due to the homogeneity of this equation, this surface is a cone with an apex at the stress-free state. The intersection of this cone with a deviatoric plane $\sigma_1 + \sigma_2 + \sigma_3 = \text{const}$ is the well-known irregular hexagon (Fig. 9.3).

Usually, the peak of a triaxial experiment with sand is used to determine the friction angle φ by Equation 9.1, and this leads to the top point of the Mohr–Coulomb hexagon in the deviatoric plane. With this φ value, the limit state for extension triaxial tests can be calculated and this determines the lowest corner of the hexagon. The remaining corners are drawn by using symmetry (or isotropy): axes 1, 2 and 3 are equivalent, and the corners are connected by straight lines (because Equation 9.1 is linear in σ_1 and σ_3). So, the entire Mohr–Coulomb limit state can in principle be inferred from *one* experiment. However, the measured peak stress

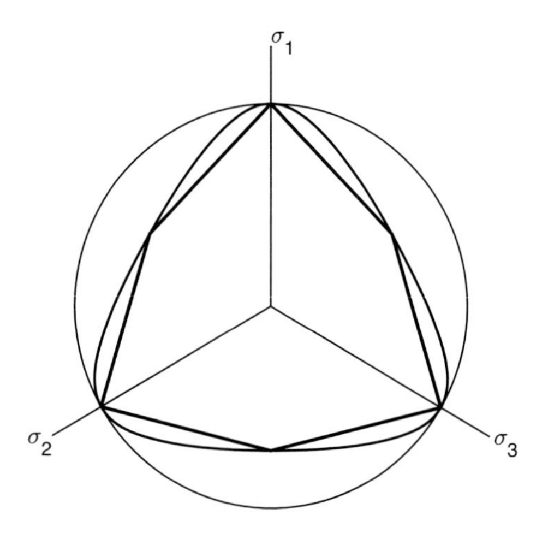

Figure 9.3 The hexagon represents in a deviatoric plane the locus of stress states fulfilling the Mohr–Coulomb limit condition for $\varphi = 30°$. The smooth curve represents the corresponding Matsuoka–Nakai limit condition and the circle represents the limit condition proposed by Drucker and Prager.

is not always reliable, because shear bands can appear before the peak of a triaxial compression and necking can occur before the peak of triaxial extension (Fig. 2.6).

The Mohr–Coulomb limit condition serves in many elastoplastic models as so-called yield surface (see Chapter 14). The related hexagon in the deviatoric plane can be approximated by a smooth curve as proposed by Matsuoka and Nakai:

$$\frac{(\sigma_1 + \sigma_2 + \sigma_3)(\sigma_1\sigma_2 + \sigma_1\sigma_3 + \sigma_2\sigma_3)}{\sigma_1\sigma_2\sigma_3} = \text{const} \tag{9.2}$$

9.3 Friction Angle

The angle of internal friction, φ, can be determined from experiments, such as triaxial or shear-box tests, where the limit state, i.e. zero stiffness, can be reached. We should distinguish between the peak value, φ_{peak}, which decreases with increasing stress level ('barotropy') and increasing void ratio ('pyknotropy'), and the critical (or residual) value, φ_c, which is a material constant for a given soil.

Barotropy and pyknotropy can be explained by relating the peak friction angle with dilatancy, according to a concept by Taylor [103], which refers to a soil sample in the shear box (Fig. 9.4): The upper half of the sample is shifted against the lower one by the horizontal displacement $\mathrm{d}s$. Due to dilatancy, the upper half of the sample is lifted from the lower half by the amount $\mathrm{d}u$. This means that a part of the work (per unit area of the sample), is used to counteract the normal stress. Thus, the incremental deformation work reads $\tau_{peak}\mathrm{d}s = \sigma\mathrm{d}u + \tau_c\mathrm{d}s$, where τ_c is the shear strength at zero dilatancy, i.e. at the so-called critical state. After division by $\sigma\mathrm{d}s$, the so-called critical friction angle φ_c is obtained:

$$\tan\varphi_{peak} = \tan\varphi_c + \mathrm{d}u/\mathrm{d}s. \tag{9.3}$$

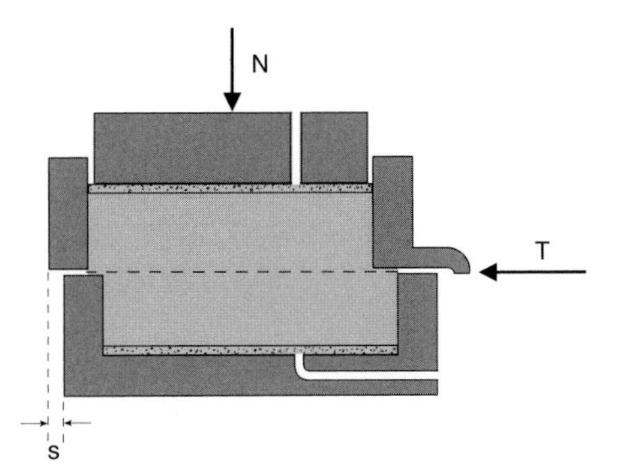

Figure 9.4 Shear box (schematically). Reproduced from [57], courtesy of Springer Nature.

Denoting du/ds as tan ψ (ψ: angle of dilatancy) we obtain:

$$\tan \varphi_{peak} = \tan \varphi_c + \tan \psi. \tag{9.4}$$

This equation explains the influence of dilatancy on the peak friction angle φ_{peak}. The dilatancy angle is pressure dependent, because the dilatancy is suppressed with increasing normal stress σ. Clearly, dilatancy depends on the void ratio e and disappears by definition at the critical void ratio e_c. Note that there are several different definitions of the dilatancy angle.

The friction angle between the individual grains has little or nothing to do with the aforementioned values φ_{peak} and φ_c.

9.3.1 Friction Angle from the Shear Box

In the shear box the two halves of a soil sample are shifted relative to each other. One would expect that the case of two rigid bodies sliding relative to each other applies best here. However, photographs with polarised light by Morgenstern & Tchalenko [73] show that the development of shear bands within the sample does not justify the concept of two rigid bodies in contact (Fig. 9.5). It still remains unclear as to whether triaxial and shear-box tests yield the same value of friction angle.

At that, the evaluation of shear-box test results is not unambiguous. With τ being the shear stress at the limit state, we usually set $\varphi = \arctan \frac{\tau}{\sigma}$. However, assuming that τ is the maximum shear stress in the shear band between the two parts of the box leads to $\hat{\varphi} = \arcsin \frac{\tau}{\sigma}$. This implies larger values of φ, e.g. for $\hat{\varphi}_2 = 30°$ we have $\varphi = 35.26°$.

9.4 Cohesion

In contrast to frictional strength, which results from normal stress σ_n, the so-called cohesion represents a strength which exists also at vanishing normal stress. With τ_f denoting the shear strength, i.e. the maximum applicable shear stress, the following equation is commonly used in soil mechanics and expresses the fact that shear strength consists of friction and cohesion:

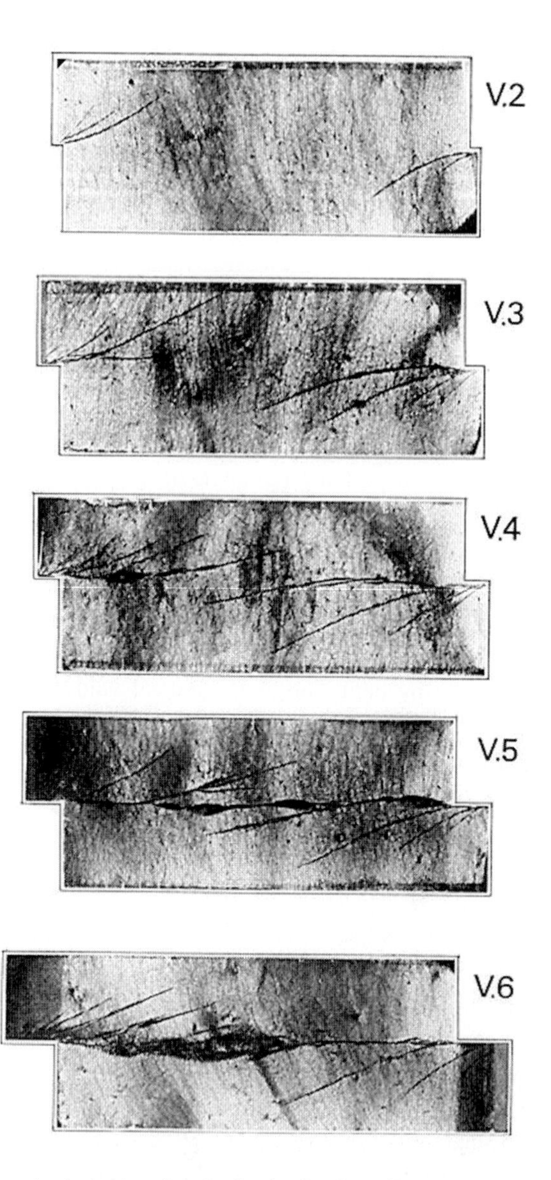

Figure 9.5 Inhomogeneous deformation in the shear-box visualised with polarised light by Morgenstern & Tchalenko. Reproduced from [73].

$$\tau_f = \sigma_n \tan \varphi + c. \tag{9.5}$$

Cohesion is a much-discussed notion in soil mechanics. On the one hand, it is important for geotechnical engineering. For instance, unsupported vertical cuts and unlined tunnels are impossible in cohesionless soils. On the other hand, measured values of cohesion exhibit a large scatter and, moreover, the physical origin of cohesion is unclear and still controversial. Some authors attribute cohesion to electromagnetic attraction between clay platelets, others say that it is nothing but the y-axis intercept of the linearisation to a non-linear friction law (Fig. 9.6) or a result of negative pore-water pressure prevailing in a dilatant water-saturated soil [92] or

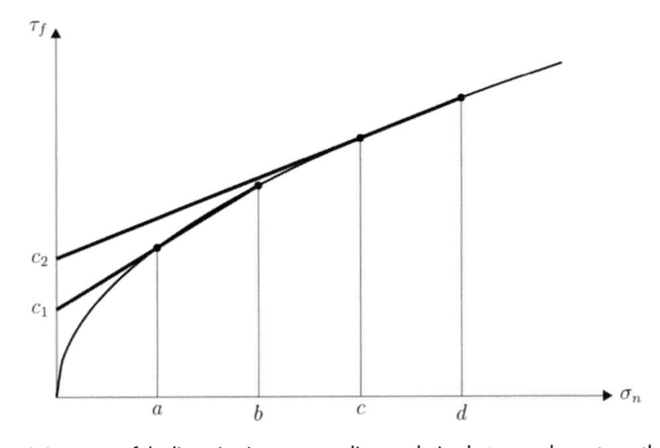

Figure 9.6 Cohesion as y-axis intercept of the linearisation to a non-linear relation between shear strength τ_f and normal stress σ_n. Linearisation in the range (a, b) leads to cohesion c_1, linearisation in the range (c, d) leads to cohesion c_2.

Figure 9.7 Folds in rock strata simulated in a sandbox. Photograph by courtesy Philip S. Prince.

due to capillary suction. The cohesion of some marine clays is attributed to dissolved salt and can be lost if this salt is leached (e.g., Norwegian quick clay).

9.5 Rock as Frictional Material

The strength of rock is also attributed to friction and cohesion. Consider a granite with a cohesion of, say, $2 \cdot 10^3$ kN/m^2, $\gamma = 20$ kN/m^3 and $\varphi = 30°$. At a depth of

$z = 10{,}000$ m, we have $\sigma = \gamma z = 2 \cdot 10^5$ kN/m^2. The shear strength by friction is $\sigma_z \tan\varphi = 1.15 \cdot 10^5$ kN/m^2. We thus see that the frictional strength is much greater than the cohesion, which implies that with respect to thick geological strata, rock can be considered as a frictional material. This is why we can simulate the behaviour of rock strata, e.g. folding (Fig. 9.7), in so-called sandbox models.

10 Collapse

10.1 Importance of Collapse in Soil Mechanics

In its early period, the principal aim of soil mechanics was to prove that a given slope or a given foundation will not fail by undergoing uncontrolled large deformations, i.e. it will not collapse. Therefore, the shear strength, as expressed by the parameters φ and c, was considered the most important property of a soil and there are a series of methods to prove whether collapse is imminent or not. These methods are typical for the early studies of soil mechanics and are characterised by the absence of considerations of the stress–strain behaviour of soil, which was unknown in any case. So, this chapter is a review of the early studies of soil mechanics.

10.2 The Phenomenon of Collapse

Strictly speaking, collapse (or 'failure') is not a singular event but rather a process that gradually evolves. However, in engineering practice, collapse is usually considered as a singular event characterised by an algebraic equation of the involved parameters. This equation is used to prove whether a construction is safe or not. In doing so, the initial boundary value problem (IBVP) consisting of the differential equations of equilibrium, the constitutive equation, the boundary and the initial conditions, is not treated, as the related difficulties are huge. Instead, approximative solutions are sought by assuming that the shear strength of the considered soil is fully mobilised within a significant region. In other words, a limit stress field is sought-after that fulfils the imposed static boundary conditions. However, this stress field need not be the solution of the IBVP, hence it is relatively easy to construct such a field (see Section 10.3). Alternatively, a so-called collapse mechanism is sought-after, i.e. a velocity field that fulfils the kinematic boundary conditions and is compatible with a fully mobilised strength. In view of the tendency of solids to localise the strain into thin shear bands (see Chapter 17), collapse mechanisms are usually considered as consisting of rigid blocks that glide relative to each other.

With both approaches, the static and the kinematic, approximations of the limit load can be obtained. To assess the safety, it would be desirable that these approximations are lower bounds of the real limit load because we would then know that the real collapse load is larger than our estimation. Besides limit loads, in geotechnical engineering, matter also support forces, e.g. the support of a soil mass, as provided by a retaining wall or a tunnel lining. A safe approximation of a support force would require an upper bound of the real support force. Such bound theorems are formulated in the realm of plasticity theory, provided the 'normality rule' (see Chapter 14) is valid. As this rule has not been confirmed for soil, these theorems are not considered here. Thus, the only thing we can do is to vary the considered stress or velocity fields in order to find a possibly low collapse load or a possibly high support force.

10.3 Plastified Zones

Estimations of collapse loads can be obtained by stress fields that fulfil the Mohr–Coulomb limit condition $\tau = \sigma_n \cdot \tan\varphi + c$. 'Construction' of such fields means here their analytical, numerical or graphical determination. In case of plane deformation, the principal stress σ_y is not considered. There are various methods for the construction of stress fields. The equilibrium equations read:

$$\frac{\partial \sigma_{xx}}{\partial x} + \frac{\partial \tau_{xz}}{\partial z} = 0, \qquad \frac{\partial \tau_{xz}}{\partial x} + \frac{\partial \sigma_{zz}}{\partial z} = 0. \tag{10.1}$$

Note that for $\gamma \neq 0$ the second equilibrium equation reads $\frac{\partial \tau_{xz}}{\partial x} + \frac{\partial \sigma_{zz}}{\partial z} = \gamma$. The substitution $\sigma_{zz} := \sigma_{zz} + \gamma z$ leads to the Equation 10.1. However, the two equilibrium equations are not sufficient to determine the three fields $\sigma_{xx}(x,z), \sigma_{zz}(x,z), \tau_{xz}(x,z)$, hence a third equation (the so-called closure equation) is needed. If one takes the constitutive equation, i.e. the stress–strain relation for the considered material, then the considered IBVP can be solved numerically, for example with finite elements. For approximate solutions the Mohr–Coulomb limit condition $\tau = \tan\varphi\ \sigma_n + c$ is used as the closure equation.

10.3.1 Slip-line Analysis, Method of Characteristics

Inserting the Mohr–Coulomb relation into the equilibrium conditions (Equation 10.1) leads to a system of two uncoupled differential equations, which is 'hyperbolic' and thus allows us to consider the so-called characteristics or slip lines. Along these lines, shear and normal stresses τ and σ_n fulfil the limit condition $\tau = c + \sigma_n \cdot \tan\varphi$. If the stress distribution is given on the boundary (this must not coincide with a slip line), the resulting stresses inside a certain region of the body can be calculated exploiting the fact that certain quantities (the so-called Riemann invariants) can be calculated from the boundary stresses and remain constant along the characteristics. The method of characteristics was applied to obtain some well-known solutions in soil mechanics [15, 29, 69, 91, 96] but is nowadays rarely used. It is presented here for completeness. For simplicity, we consider the special case of a purely cohesive material ($c > 0, \varphi = 0$).

With $\sigma := (\sigma_1 + \sigma_2)/2$, $\tau := (\sigma_1 - \sigma_2)/2$, $\cos 2\zeta = -\sin 2\vartheta$ and $\sin 2\zeta = \cos 2\vartheta$ (Fig. 10.1) we have:

$$\begin{aligned} \sigma_x &= \sigma + \tau\cos 2\zeta = \sigma - \tau \sin 2\vartheta \\ \sigma_z &= \sigma - \tau\cos 2\zeta = \sigma + \tau \sin 2\vartheta \\ \tau_{xz} &= \tau \sin 2\zeta = \tau\cos 2\vartheta. \end{aligned} \tag{10.2}$$

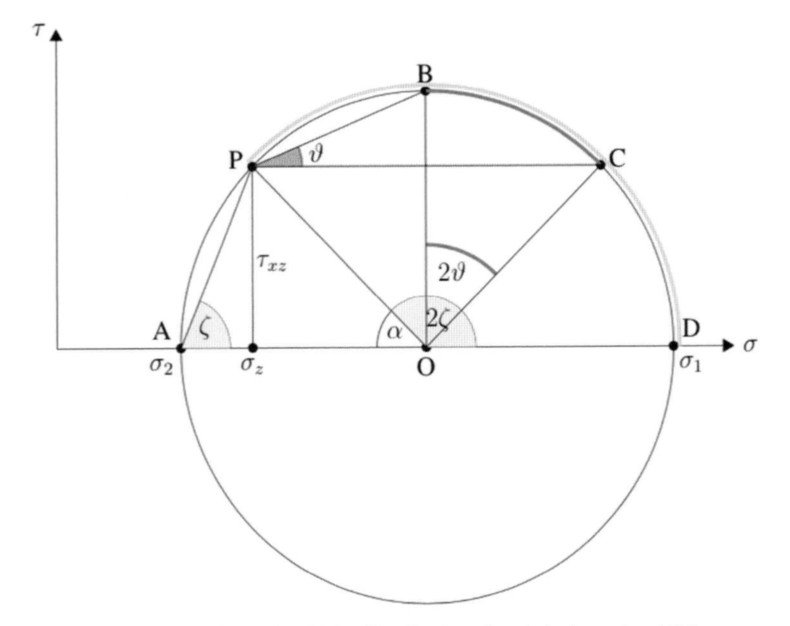

Figure 10.1 P is the pole of the Mohr circle. The angles PAD ($= \zeta$, inclination of a principal stress) and POD are peripheral and central angles. The same holds for the angles BPC ($= \vartheta$, inclination of a slip line) and BOC. Obviously, $\alpha = 90^\circ - 2\vartheta = 180^\circ - 2\zeta$, hence, $2\zeta = 90^\circ + 2\vartheta$.

Fulfilment of the limit condition $\tau = c$ implies that the stress state is now given by the variables σ and ϑ (σ_y is not considered here). Introducing $\tau = c$ and Equation 10.2 into Equation 10.1 yields:

$$\frac{\partial \sigma}{\partial x} - 2c\left(\cos 2\vartheta \frac{\partial \vartheta}{\partial x} + \sin 2\vartheta \frac{\partial \vartheta}{\partial z}\right) = 0, \tag{10.3}$$

$$\frac{\partial \sigma}{\partial z} - 2c\left(\sin 2\vartheta \frac{\partial \vartheta}{\partial x} - \cos 2\vartheta \frac{\partial \vartheta}{\partial z}\right) = 0. \tag{10.4}$$

Equations 10.3 and 10.4 form a system of coupled partial differential equations for $\sigma(x,z)$ and $\vartheta(x,z)$. It has the form

$$a_{11}\frac{\partial \sigma}{\partial x} + a_{12}\frac{\partial \vartheta}{\partial x} + b_{11}\frac{\partial \sigma}{\partial z} + b_{12}\frac{\partial \vartheta}{\partial z} = 0$$

$$a_{21}\frac{\partial \sigma}{\partial x} + a_{22}\frac{\partial \vartheta}{\partial x} + b_{21}\frac{\partial \sigma}{\partial z} + b_{22}\frac{\partial \vartheta}{\partial z} = 0,$$

with $a_{11} = 1,\ a_{21} = 0,\ a_{12} = -2c\cos 2\vartheta,\ a_{22} = -2c\sin 2\vartheta, b_{11} = 0,\ b_{21} = 1,\ b_{12} = -2c\sin 2\vartheta,\ b_{22} = 2c\cos 2\vartheta$. With

$$\mathbf{A} = \begin{pmatrix} a_{11} & a_{12} \\ a_{21} & a_{22} \end{pmatrix}, \quad \mathbf{B} = \begin{pmatrix} b_{11} & b_{12} \\ b_{21} & b_{22} \end{pmatrix}, \quad \mathbf{t} = \begin{pmatrix} \sigma \\ \vartheta \end{pmatrix},$$

it can be written as

$$\mathbf{A}\frac{\partial \mathbf{t}}{\partial x} + \mathbf{B}\frac{\partial \mathbf{t}}{\partial z} = \mathbf{0}. \tag{10.5}$$

To determine the unknowns $\frac{\partial \sigma}{\partial x}$, $\frac{\partial \sigma}{\partial z}$, $\frac{\partial \vartheta}{\partial x}$, $\frac{\partial \vartheta}{\partial z}$, we add another two equations by considering the total differentials

$$\frac{\partial \sigma}{\partial x}\mathrm{d}x + \frac{\partial \sigma}{\partial z}\mathrm{d}z = \mathrm{d}\sigma, \tag{10.6}$$

$$\frac{\partial \vartheta}{\partial x}\mathrm{d}x + \frac{\partial \vartheta}{\partial z}\mathrm{d}z = \mathrm{d}\vartheta. \tag{10.7}$$

The so-called Cauchy problem is: Suppose we know the stress distribution on a curve k. Can we determine the stress at adjacent points? In other words, we want to know how does the stress state propagate laterally from k. The problem is solvable if we know $\frac{\partial \mathbf{t}}{\partial x}$ and $\frac{\partial \mathbf{t}}{\partial z}$ on the curve k. To solve this system of, in total, four scalar equations (Equations 10.5 and 10.6), we consider its determinant D:

$$D := \begin{vmatrix} a_{11} & a_{12} & b_{11} & b_{12} \\ a_{21} & a_{22} & b_{21} & b_{22} \\ \mathrm{d}x & 0 & \mathrm{d}z & 0 \\ 0 & \mathrm{d}x & 0 & \mathrm{d}z \end{vmatrix} = \begin{vmatrix} 1 & -2c\cos 2\vartheta & 0 & -2c\sin 2\vartheta \\ 0 & -2c\sin 2\vartheta & 1 & 2c\cos 2\vartheta \\ \mathrm{d}x & 0 & \mathrm{d}z & 0 \\ 0 & \mathrm{d}x & 0 & \mathrm{d}z \end{vmatrix} \tag{10.8}$$

Vanishing of D determines curves, the so-called characteristics of Equation 10.5, for which the Cauchy problem has no solution. Thus,

$$D = 2c\left[(\mathrm{d}x)^2 \sin 2\vartheta - 2(\mathrm{d}x)(\mathrm{d}z)\cos 2\vartheta - (\mathrm{d}z)^2 \sin 2\vartheta\right] = 0 \tag{10.9}$$

is a quadratic equation for $z' := \mathrm{d}z/\mathrm{d}x$ and has two real solutions:

$$z' = \frac{\mathrm{d}z}{\mathrm{d}x} = \left\{ \begin{array}{c} \tan\vartheta \\ -\cot\vartheta \end{array} \right\}, \tag{10.10}$$

which means that the system of differential equations is hyperbolic. We thus get two families of characteristics (or slip lines) with slope of either ϑ or perpendicular to it (i.e. for the case $\varphi = 0$ the characteristics intersect at a right angle).

Solubility of the system of four linear equations further requires that the rank of the extended matrix must be equal to the rank of (Equation 10.8), i.e.

$$\mathrm{rank}\begin{pmatrix} \mathbf{A} & \mathbf{B} \\ \mathrm{d}x\mathbf{1} & \mathrm{d}z\mathbf{1} \end{pmatrix} = \mathrm{rank}\begin{pmatrix} \mathbf{A} & \mathbf{B} & \mathbf{0} \\ \mathrm{d}x\mathbf{1} & \mathrm{d}z\mathbf{1} & \mathrm{d}\mathbf{t} \end{pmatrix}, \tag{10.11}$$

i.e., $\mathbf{t}$ may not take arbitrary values on the characteristics. We obtain the relation on the characteristics, by replacing any column of the determinant D in Equation 10.8 by the column vector $\binom{\mathbf{0}}{\mathrm{d}\mathbf{t}}$ and then set $D = 0$:

$$\begin{vmatrix} 1 & -2c\cos 2\vartheta & 0 & 0 \\ 0 & -2c\sin 2\vartheta & 1 & 0 \\ \mathrm{d}x & 0 & \mathrm{d}z & \mathrm{d}\sigma \\ 0 & \mathrm{d}x & 0 & \mathrm{d}\vartheta \end{vmatrix} = \mathrm{d}\vartheta \begin{vmatrix} 1 & -2c\cos 2\vartheta & 0 \\ 0 & -2c\sin 2\vartheta & 1 \\ \mathrm{d}x & 0 & \mathrm{d}z \end{vmatrix} - \mathrm{d}\sigma \begin{vmatrix} 1 & -2c\cos 2\vartheta & 0 \\ 0 & -2c\sin 2\vartheta & 1 \\ 0 & \mathrm{d}x & 0 \end{vmatrix}$$

$$= -2c\,\mathrm{d}\vartheta(\sin 2\vartheta \mathrm{d}z + \cos 2\vartheta \mathrm{d}x) + \mathrm{d}\sigma \mathrm{d}x$$

$$= 0.$$

Dividing by $\mathrm{d}x$ gives: $\mathrm{d}\sigma = 2c(\sin 2\vartheta \cdot z' + \cos 2\vartheta)\mathrm{d}\vartheta$. For $z' = \tan\vartheta$ we obtain (with $\sin 2\vartheta \cdot \tan\vartheta + \cos 2\vartheta \equiv 1$) $\mathrm{d}\sigma = 2c\,\mathrm{d}\vartheta$, and for $z' = -\cot\vartheta$ we obtain $\mathrm{d}\sigma = -2c\,\mathrm{d}\vartheta$. Thus, on the one family of characteristics $\mathrm{d}(\sigma - 2c\vartheta) = 0$ holds, i.e. $\sigma - 2c\vartheta$ remains constant. On the other family $\mathrm{d}(\sigma + 2c\vartheta) = 0$ holds, i.e. $\sigma + 2c\vartheta$ remains constant. It follows that one may not apply arbitrary stress distributions to characteristics (or slip lines) but constant values for $\sigma - 2c\vartheta$ and $\sigma + 2c\vartheta$.

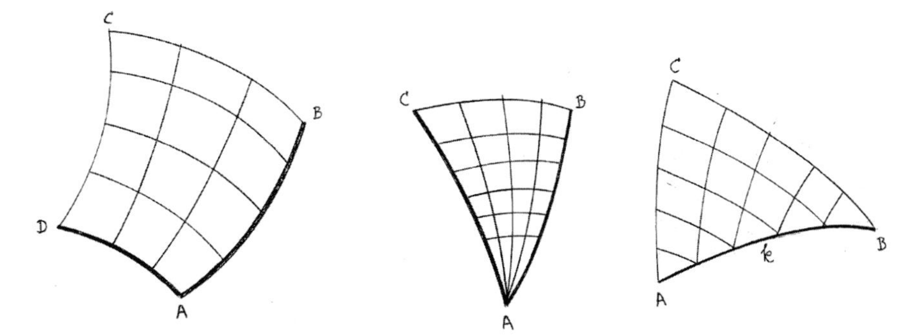

Figure 10.2 Riemann and Cauchy problems. Starting from the boundaries, the stress state can be determined in the interior of the areas ABCD or ABC.

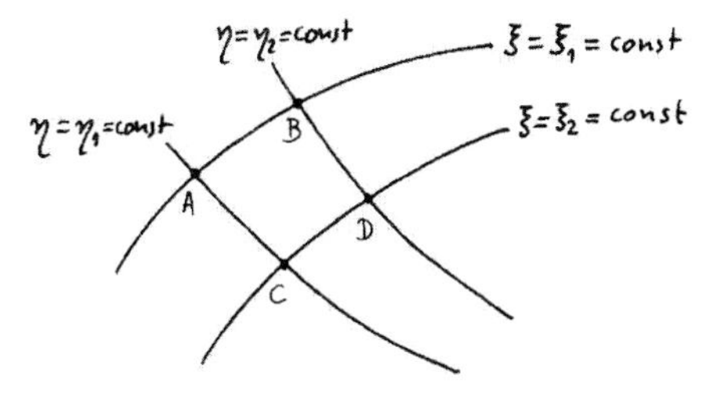

Figure 10.3 Homologous points on slip lines.

In the case of the Riemann problem, the quantity $\sigma - 2c\vartheta$ = const is given on the slip line AB (Fig. 10.2) and the quantity $(\sigma + 2c\vartheta)$ = const is given on AD. Evaluating the invariants, the stress state can be determined at each point of the region ABCD. For a centred fan of characteristics, the line AD shrinks into the point A. In the case of the Cauchy problem, the stress is known along the curve k and can be determined in the area ABC.

Let us consider two pairs of slip lines $\xi = \frac{\sigma}{2c} - \vartheta = \text{const}$ and $\eta = \frac{\sigma}{2c} + \vartheta = \text{const}$ (Fig. 10.3). Along slip lines of the one family, the other family defines homologous points. From $\sigma_A = c(\xi_1 + \eta_1)$, $\vartheta_A = \frac{1}{2}(\eta_1 - \xi_1)$, $\sigma_B = c(\xi_1 + \eta_2)$, etc. follows:

$$\sigma_A - \sigma_B = \sigma_C - \sigma_D, \quad \vartheta_A - \vartheta_B = \vartheta_C - \vartheta_D$$
$$\sigma_A - \sigma_C = \sigma_B - \sigma_D, \quad \vartheta_A - \vartheta_C = \vartheta_B - \vartheta_D.$$

This means that the differences of σ and ϑ values between homologous points remain constant along the characteristics. If in an area the slip lines are straight lines ($\vartheta \equiv$ const), then $\sigma \equiv$ const, i.e. the stress is constant (or 'uniform') in this area.

As an example, consider a load strip on a weightless ($\gamma = 0$) and frictionless ($\varphi = 0$) halfspace of cohesion c. We search the maximum applicable vertical stress p assuming that in the soil below the load strip the strength is fully mobilised, i.e. $\sigma_1 - \sigma_2 = 2c$. To this end we consider the slip line field shown in Fig. 10.4. In particular, we consider the slip line that connects the points I and F. At point I we have $\sigma_z = 0 = \sigma_2$, and the maximum principal stress σ_1 is obtained from $\sigma_1 - \sigma_2 = 2c$ as $\sigma_1 = 2c$. Similarly, at point F we have $\sigma_1 = p$ and $\sigma_2 = p - 2c$, consequently $\sigma = \frac{p+p-2c}{2} = p - c$. Taking into account that the quantity $\sigma - 2c\vartheta$ remains constant along the curve IHGF,

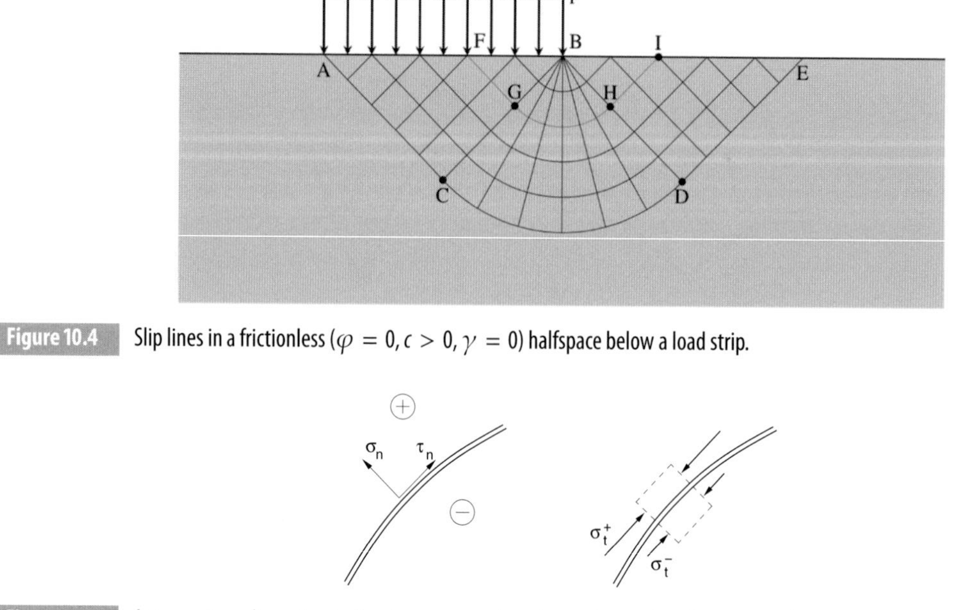

Figure 10.4 Slip lines in a frictionless ($\varphi = 0, c > 0, \gamma = 0$) halfspace below a load strip.

Figure 10.5 Across a stress discontinuity the stresses τ_n and σ_n do not change, but the tangential stress σ_t can suffer a jump. Reproduced from [57], courtesy of Springer Nature.

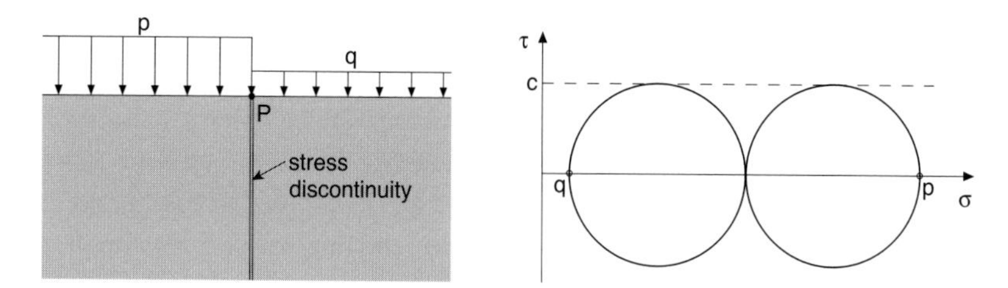

Figure 10.6 Stress discontinuity below the edge of a load strip. Reproduced from [57], courtesy of Springer Nature.

i.e. $\sigma_F - 2c\vartheta_F = \sigma_I - 2c\vartheta_I$, we obtain with $\vartheta_F - \vartheta_I = \pi/2$ the limit load (solution of Prandtl) as:

$$p = (\pi + 2)c \approx 5.14\ c. \qquad (10.12)$$

10.3.2 Stress Discontinuities

Stress discontinuities allow a relatively simple construction of stress fields. Across a stress discontinuity, the shear stress τ_n and the normal stress σ_n must not suffer a jump, whereas the corresponding tangential stress σ_t can jump (Fig. 10.5).

A simple application results in the case of a strip load p, which acts on a frictionless and weightless cohesive soil (Fig. 10.6). Outside the strip acts a lower vertical stress q. If one considers a vertical stress discontinuity by the point P and further assumes that both, below p and below q, limit states prevail, it follows that below load q the smallest principal stress is q and, consequently, the largest principal stress is $q + 2c$. Stress equilibrium across the stress discontinuity requires that the horizontal normal stresses be equal in both regions. This implies that the minimum principal stress in this region, $p - 2c$, be equal to $q + 2c$. Thus, the estimation of the collapse load is obtained as

$$p = q + 4c. \tag{10.13}$$

With $p = 4c$ (for $q = 0$) we thus obtain a lower estimate than the one given by Equation 10.12. Slightly higher values for p can be obtained if more than one stress discontinuity is set at the point P.

10.4 Collapse Mechanisms

A collapse mechanism is a velocity field that prevails at collapse. It need not be *the* actual velocity field that prevails at collapse. It is rather a possible velocity field, i.e. a kinematic concept of collapse. According to observations, realistic collapse mechanisms consist of rigid blocks that glide relative to each other. Consequently, the boundaries of these blocks must be surfaces that can shift in themselves, i.e. planar or cylindrical surfaces. So, in two-dimensional (2D) problems either straight lines or circles.

10.4.1 Blocks with Circular Slip Surfaces

The example of Fig. 10.7 shows a failure of the soil (assumed as cohesive but frictionless and weightless) underneath a load strip. Failure sets in as soon as the moment exerted by the cohesion along the circular slip line cannot resist the driving moment exerted by the load p. The geometry of this collapse mechanism is represented by the angle α and the radius $r = b/\sin(\alpha/2)$ yielding the resisting moment $M_R = crl = cr^2\alpha$, whereas the driving moment is $M_D = \frac{1}{2}pb^2$. The ratio $\eta := \frac{M_R}{M_D} = \frac{4c}{p}\frac{\alpha/2}{\sin^2(\alpha/2)}$ is minimum for $\alpha \approx 132.6°$ (this results from $\mathrm{d}\eta/\mathrm{d}\alpha = 0$). Hence,

$$p = 5.5\,c. \tag{10.14}$$

The comparison of resisting and driving moments becomes difficult for circular slip surfaces in soils with $\varphi > 0$, because in this case we need the normal stress distribution along the slip surfaces, which is unknown however and, thus, the moments cannot be figured out. As a way out, one has tried to divide the sliding block into vertical strips assuming that the weight of each strip acts only downwards (Fig. 10.8). Of course, this is but a poorly justified assumption. Improved solutions were sought with various assumptions about the forces between the strips. These attempts rather

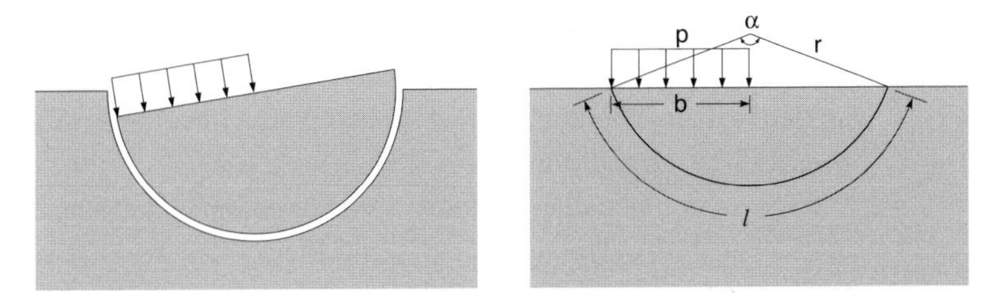

Figure 10.7 Circular collapse mechanism for a load strip on cohesive soil with $\varphi = 0, \gamma = 0$. Reproduced from [57], courtesy of Springer Nature.

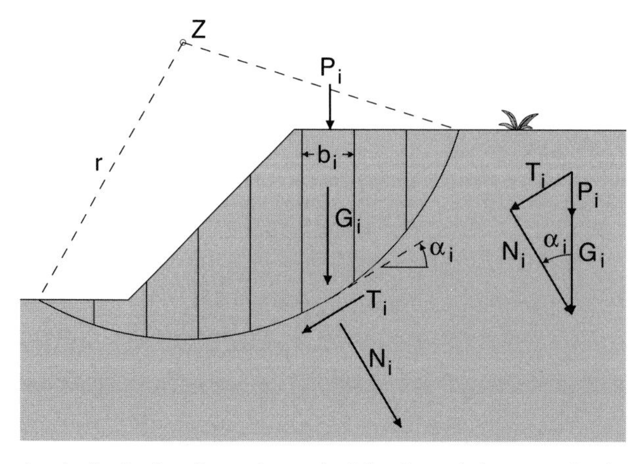

Figure 10.8 In order to determine the distribution of normal stress (and thus friction) along a circular slip surface, the Fellenius method assumes that the weight of each strip acts only downwards. Reproduced from [57], courtesy of Springer Nature.

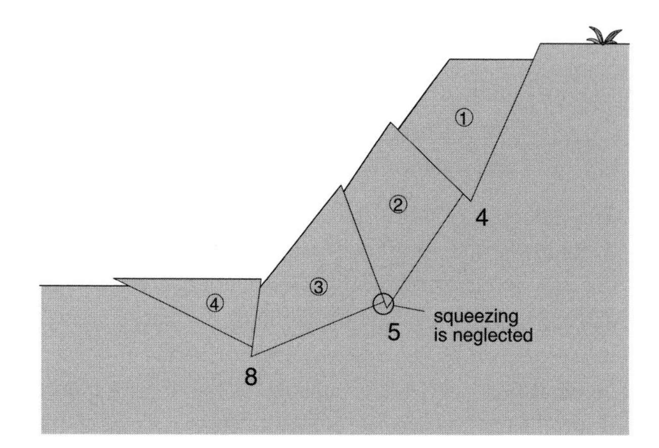

Figure 10.9 Collapse mechanism for a slope. It consists of four blocks. The displaced positions of the blocks are depicted. The squeezing of the edges is neglected. Reproduced from [57], courtesy of Springer Nature.

mirror the widespread but not justified expectation that increasing complexity yields better results.

10.4.2 Blocks with Planar Slip surfaces

In collapse mechanisms with planar slip surfaces, the individual blocks undergo translations but no rotations. Consequently, moments cannot be considered. Nevertheless, translative collapse mechanisms prove to be a versatile tool that can be easily applied to many 2D situations for the estimation of collapse loads. The generalisation towards 3D is straightforward.

To estimate the collapse load, a realistic collapse mechanism should first be laid down (Fig. 10.9). The number of rigid blocks can be kept small. Contrary to the method of finite elements, a mesh refinement does not bring any substantial benefit. Clearly, the geometry of the mechanism is arbitrary and has to be varied so as to

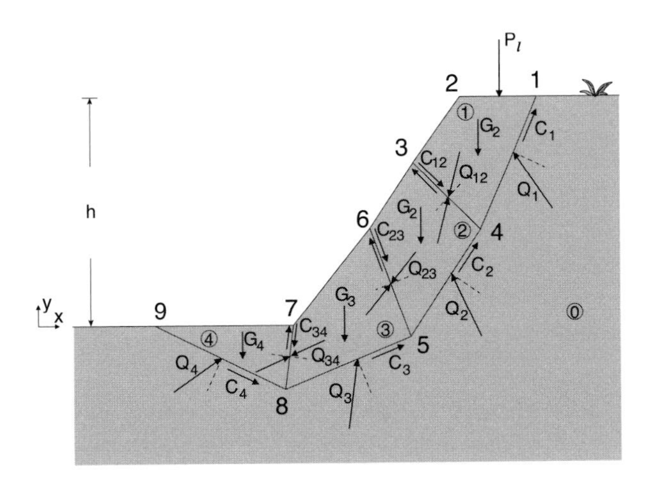

Figure 10.10 Forces acting upon each block of the collapse mechanism of Fig. 10.9. Reproduced from [57], courtesy of Springer Nature.

provide the minimum collapse load. Next, the displacement of the individual rigid blocks will be considered. Instead of 'displacement' we can use the word 'velocity'. Denoting each block by a number i and the immobile part of the subsoil with 0 we introduce the relative velocities by

$$\mathbf{v}_{ij} = \mathbf{v}_i - \mathbf{v}_j, \tag{10.15}$$

where $\mathbf{v}_{ij}$ is the velocity of block i, as observed from block j, and $\mathbf{v}_i = \mathbf{v}_{i0}$, etc. Writing down Equation 10.15 for all blocks yields a system of linear equations to determine the velocities of each block. We usually choose so-called single kinematic chains, so that prescribing a single velocity component suffices to determine all other components. If no solution can be obtained, then the mechanism is self-locking and should be rejected. In the 2D case, the solution can be obtained graphically. The here presented mathematical evaluation can be easily applied in the 3D case.

The absolute value of the displacements is immaterial, as only their relations count. Clearly, the displacements cause squeezing of the corners, but this is neglected. The cohesion force $\mathbf{C}_{ij}=cA_{ij}\mathbf{v}^0_{ij}$ is the force with which block i acts upon the adjacent block j due to the cohesion c acting in the mutual contact surface A_{ij}. Block i acts upon the adjacent block j also by the frictional force $\mathbf{Q}_{ij} = Q_{ij}(\mathbf{n}_{ij} - \mathbf{v}^0_{ij} \tan\varphi)$, where $\mathbf{n}_{ij}$ is the unit normal vector pointing outwards of block i on the joint surface with block j, and the unit vector $\mathbf{v}^0_{ij}$ is defined as $\mathbf{v}_{ij}/|\mathbf{v}_{ij}|$. The coefficients Q_{ij} can be obtained by establishing equilibrium at each block taking also into account its weight $\mathbf{G}_i$, see Fig. 10.10. In this way, the external force $\mathbf{P}_l$ that causes collapse (i.e. the limit load) can be determined. The geometry of the collapse mechanism should be varied to obtain the minimum collapse load.

10.4.3 Coulomb's Estimation of the Active Earth Pressure

Considering a collapse mechanism consisting of a single rigid block that glides along a plane inclined by the angle ϑ, Coulomb estimated the active earth pressure acting upon a retaining wall. He considered collapse induced by the yielding of the retaining

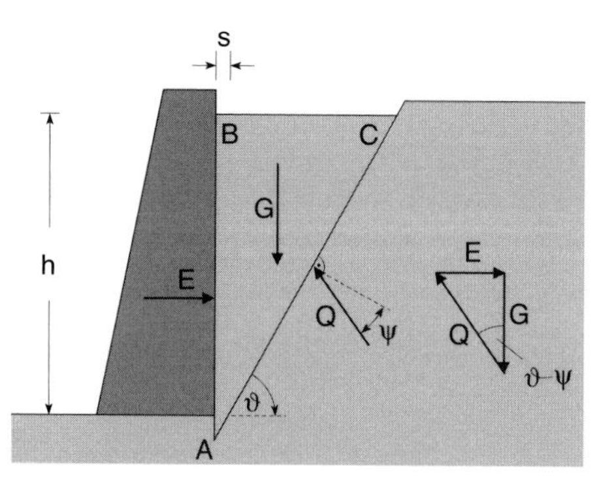

Figure 10.11 Collapse mechanism of Coulomb for the active earth pressure. Here the wall friction angle is assumed to $\delta = 0$ and, therefore, the earth pressure E is perpendicular to the line AB. Reproduced from [57], courtesy of Springer Nature.

wall, i.e. a displacament to the left by a non-further specified value s (Fig. 10.11). As a consequence, a prismatic soil block slides along a slip plane inclined by ϑ. The value of the wall friction angle has to be assumed. Applying calculus, he varied ϑ to find the minimum value of the earth pressure E. Setting $\mathrm{d}E/\mathrm{d}\vartheta = 0$ is sometimes considered as an additional principle introduced by Coulomb. However, this is not the case, because this equation is implied by the fact that imminent collapse is considered [57]. A similar approach can be applied to determine the passive earth pressure. Active and passive earth pressures are but two limits that bound the real value of the earth pressure, which is hard to predict. This is to be contrasted with the exact prediction of water pressure.

10.5 Safety

Assuming that collapse is a singular event, the engineer wants to show that a geotechnical situation (e.g. a cut) is far enough from a collapse. He calls this distance 'safety' (or 'factor of safety') and wants to prove that a considered structure has sufficient safety. But how to define this distance and how to prove that a particular safety is sufficient? These questions are not yet answered. Moreover, they cannot be answered in an objective way. Let us consider them separately.

Safety is a sort of distance in the event space. Assuming that this space has distances, how can we measure the distance between a given state and collapse? There is no unique answer to this question, as a distance (or a metric) can be introduced in many different ways. An objective measure for distance could perhaps be the probability of collapse. In fact, much effort has been spent towards a probabilistic approach to collapse. However, it remained inconclusive, mainly because of lack of data sufficient to introduce statistical considerations. It should be added that the probability of collapse cannot be estimated even for structures composed of industrially manufactured parts, e.g. a space flight [24]. It should also be taken into account that the collapse (or failure) probability is a – possibly – very small number, but still >0. In other words, there is always a residual risk. Hence, the question arises on how safe is safe enough? This question refers to the acceptance of residual risk. Its answer is subjective. Individuals tend to answer by imitation thinking, e.g. 'if my friend accepts the risk of an air travel, then I am also ready to fly'. Of course, this readiness may vary from country to country, say if the aeroplanes of country X are less safe than the ones of country Y. In other words, the question on a (whatsoever measured) safety and acceptable residual risk has different answers in countries with different traditions, authorities and levels of technology. This is why the huge efforts on harmonisation of national technical codes on safety have failed so far.

10.5.1 Probabilistic Approach to Safety

The sources of failures in geotechnical engineering are manifold [35]:

1. Uncertainties in the load assumptions.
2. Insufficient knowledge of the material behaviour (constitutive equation).
3. Spatial variation of soil properties.
4. Measurement errors in field and laboratory tests.
5. Error in the solution of the relevant boundary value problem (e.g. as a result of numerical approximation).
6. Defects of design and execution.
7. Ageing, weathering and wear of the individual supporting elements.

Sources 1, 2, 5 and 7 are essentially based on modelling and can be remedied in principle (e.g. through advances in research), whereas the other errors are stochastic. Let us assume that the process of model building is already completed and let us consider the safety in view of the stochastic errors. We consider the relevant variables of a foundation or of a construction as random variables x_i. The limit state is given by the limit state equation

$$g(x_1, x_2, x_3, \ldots) = 0$$

so that points x_i with $g(x_i) > 0$ are safe states and points with $g(x_i) < 0$ represent unsafe states. The probability of failure P_f is then given by

$$P_f = \int_{\mathcal{B}} f(x_1, x_2, x_3, \ldots, x_n)\, \mathrm{d}x_1 \mathrm{d}x_2 \ldots \mathrm{d}x_n, \tag{10.16}$$

where the region $\mathcal{B}$ of the x_i-space is characterised by the inequality $g(x_i) < 0$ and $f(.)$ is the joint probability density function. The probability of failure P_f, or the reliability $\mathfrak{R} = 1 - P_f$, should be calculated from the multiple integral (Equation 10.16) (so-called level III safety proof) but this is rather academic. Apart from the fact that the density function is unknown, the calculation of the multiple integral is difficult. One therefore tries to find a so-called safety proof of level II and considers the random variable

$$z := g(x_1, x_2, \ldots, x_n). \tag{10.17}$$

Its mean η_z divided by the standard deviation σ_z,

$$\beta := \frac{\eta_z}{\sigma_z},$$

can be seen as a measure of the probability of failure $P\{z < 0\}$, which should be sufficiently small. Therefore, the so-called safety index β must be sufficiently large. The question now arises to what extent the probability of failure can be calculated from β. In the case that

- the random variables x_i are normally distributed (a density function can hardly be determined by experiments) and
- $g()$ is a linear function, i.e.

$$z = c_0 + c_1 x_1 + c_2 x_2 + \cdots + c_n x_n, \tag{10.18}$$

(for approximate calculations, the requirement for normal distribution of the variables x_i need not be strictly fulfilled, since $z = c_0 + c_1 x_1 + c_2 x_2 + \cdots$ is approximately normally distributed by virtue of the central limit theorem),

it can be shown that the probability of failure is

$$P_f = \Phi(-\beta) = \frac{1}{\sqrt{2\pi}} \int_{-\infty}^{-\beta} e^{-x^2/2} \mathrm{d}x.$$

In this case, β can be calculated as follows:

$$\eta_z = E\{z\} = c_0 + c_1 E\{x_i\} + c_2 E\{x_2\} + \cdots + c_n E\{x_n\},$$

$$\sigma_z = \sqrt{c_1^2 \sigma_1^2 + c_2^2 \sigma_2^2 + \cdots + c_n^2 \sigma_n^2}$$

with $E\{x_i\}$ being the expected value of x_i. Hence,

$$\beta = \frac{\eta_z}{\sigma_z}.$$

If the limit state equation is non-linear, it should be linearised

$$g(x_i) \approx g(x_{i0}) + \nabla g \mid_{x_{i0}} \cdot (x_i - x_{i0}).$$

The linearisation occurs at point x_{i0}, the so-called design point x_1^* should be chosen. This is the most probable failure state, i.e. the most probable point on the hypersurface $g(x_i) = 0$. The design point therefore results as the solution of the extremal problem $f(x_i) = \max$ with the constraint $g(x_i) = 0$. According to Lagrange, this point is characterised by the fact that an isoline (or hyperplane) $g(x_i) = \text{const}$ is tangent to an isoline $f(x_i) = \text{const}$, respectively $\nabla f = \lambda \nabla g$ holds.

The linearisation around the design point has, among other things, the consequence that the function $g(x_i)$ becomes linearly homogeneous in the sense that $g(\lambda x_i) = \lambda g(x_i)$ holds. This makes β invariant to changes in the dimension of the variables x_i as well as to algebraic transformations of the limit state equation $g(x_i) = 0$.

10.5.2 Harmonisation of Codes

When it came to harmonising European codes and in particular the individual views on safety, it was realised that the common definition of safety is not a physical quantity and that the only objective term for it is the probability of failure. The problem, however, is that the probability of failure cannot be expressed objectively either but can at best be estimated with a probability. But is this really possible?

For a long time there was a rumour that NASA could calculate the failure probability of a space flight from the failure probability of each individual screw. However, on the occasion of the Challenger disaster on 28 January 1986, the Nobel Prize winner R. Feynman discovered that this was by no means the case [24].

Also, geotechnical engineers have had to realise that a probabilistic definition of safety is not feasible. Initially, they managed well as long as they worked in the closed environments of their countries where their concepts of safety, however developed by local authorities, had become established and had proved themselves well. The problems only came to light when there was an attempt to harmonise these concepts at an international level. Quite similar to a religion, with which one can live quite comfortably as long as one does not come into contact with foreign peoples who have other gods. However attractive the word harmonisation sounds; its realisation proves problematic. Imagine what harmonisation would result in if, for example, the United Nations decided to harmonise all the religions of the world!

10.6 Imminent Collapse

How likely is an existing soil or rock structure, e.g. a steep slope, to collapse? The answer to this question is extremely difficult and may be linked to costly protective measures. One possible answer would be to take action when the (factor of) safety is only slightly higher than 1. Taken, however, that the evaluation (let alone the measurement) of safety is not possible, it only remains to proceed intuitively taking into account the possible exposure of sensitive structures and also existing evidence of rockfalls (Fig. 10.12). At any rate it should be taken into account that the mere existence of a structure is no guarantee that it will remain standing in the future.

Figure 10.12 Steep rock slopes with different susceptibilities to rockfall. (a) near Sydney, the debris testifies to permanent break-off. (b) Meteora, Varlaam monastery, Greece. Photograph: Danimir, Wikimedia Commons.

11 Constitutive Equations

11.1 Constitutive Equation versus Constitutive Law

Constitutive equations describe the mechanical behaviour of a material. Alternative expressions are ‘constitutive relation’ or ‘constitutive law’, and the choice between them is a matter of preference. Whereby, the designation ‘constitutive law’ has a more binding character. To cite Truesdell [111], ‘The status of a “Law” of physics: a clear, precise concept of ideal behavior, embracing an enormous variety of precisely specifiable cases’.

11.2 Why Do We Need Constitutive Equations?

The constitutive equation of soil is of central importance in soil mechanics. Nevertheless, many geotechnical engineers do not know any constitutive equations for soil. Moreover, they do not know why they are required. Usually, they have to prove that failure will not occur. To this end, they use calculation methods based on friction angle φ and cohesion c and believe that they can thus capture the behaviour of soil. As for the deformations of the ground, they have learned that these can hardly be captured by calculations. For demanding numerical simulations, computer programs with built-in complex and hardly tractable constitutive laws are applied. By and large, people have learned to work without a constitutive equation for soil. In their view, there is no urgent economic need for it.

So, why should we, therefore, still need research on constitutive equations for soil? – for two reasons. First, the present accuracy of numerical simulations is not satisfactory (see Chapter 20). Moreover, there are still unanswered problems, whose treatment requires improved tools. The second reason is of a different nature and, perhaps, more important. It is an intellectual challenge to understand the behaviour of granular materials, which are pretty diverse and comprise the mostly of our food sources (wheat, sugar, etc.) as well as the ground upon which we live. Their behaviour is extremely complex (see Chapter 2) and can only be understood on the basis of a convincing constitutive relation. To aid understanding, the equation we are looking for must be *simple*.

11.3 Are Constitutive Equations Dispensable in View of Artificial Intelligence?

This question can be answered on the basis of the planetary motion and Newton's law of mass attraction. This law allows the planetary orbits to be calculated exactly and is based on astronomical observations by Tycho Brahe, which led Kepler to his laws, and which were then sublimated into Newton's law of mass attraction. With today's artificial intelligence (AI), Kepler's and Newton's achievements would be dispensable, we would feed our computers with Tycho Brahe's data and they would calculate the planetary orbits. However, the understanding of planetary motion would be missing.

An important question arises from this: Should we search for constitutive equations at all or should we be content with the conclusions drawn from an artificial or other kind of empirical intelligence? The question is justified if one considers that the belief in the utility of such laws is nowadays shattered. In addition, there is the everlasting hostility of practitioners against theory as well as the blind trust in powerful computers.

A remarkable attempt towards this direction is to carry out numerical simulations, and instead of constitutive equations, to use the results of numerical simulations with distinct elements [20].

However, we cannot interact with the world and nature as *tabulae rasae*, even if we have great computers. If computers are to help us, we must address them with meaningful questions. They also have to be fed with meaningful data. So we need meaningful concepts, and these can only be formed within the framework of meaningful theories. Theories do not follow directly from nature but result from the creative human mind. According to Kant (cited in Popper) 'The mind does not draw its laws ... from nature, but prescribes them to it' [85].

11.4 Material Constants

In every constitutive equation appear some material constants that characterise a particular material, e.g. in Hooke's law Youngs's modulus E and Poisson's ratio ν appear. Thus, material constants are mere numbers that characterise a particular material *in the realm of a particular constitutive relation.*

The often-posed question on the physical meaning of the material constants is not reasonable. It only makes sense for cases where a single material constant happens to influence the outcomes of a particular experiment.

The widespread attitude in soil mechanics to characterise material constants as 'drained' and 'undrained' is not reasonable. The same holds for characterising them as dynamic, e.g. dynamic Young's modulus.

11.5 Calibration

Using a constitutive equation, one should be able to simulate element tests, i.e. to predict their results. To obtain this, the constitutive equation must be adjusted to a particular material. This is done by determining the material constants, which enter as free parameters into the constitutive equation. If one considers the numerical simulation of element tests as the 'direct' problem, then the calibration of a constitutive law, i.e. the numerical determination of the material constants based on results, represents a so-called *inverse problem*. Inverse problems are often characterised by the fact that small variations of the input data can cause large variation in the results. Thus, they are not 'well-posed' problems (see Chapter 17). The difficulty of calibration increases disproportionately with the complexity of a constitutive equation, and there is no lack of constitutive equations that are almost impossible to calibrate (and thus virtually useless).

Calibration is an interpolation (or approximation) task of often underestimated difficulty which arises for two reasons:

1. The input data are burdened not only by natural scatter but also by unknown systematic errors. At that, the interpolating function (i.e. the constitutive relation) does not have absolute validity. This is to be contrasted, e.g. with the determination by Gauss of a planetary orbit on the basis of measurements. In this case, we know exactly that the orbit is an ellipse and, therefore, a least square fit leads to a unique result.
2. Contrary to a widespread expectation, a constitutive relation is not a collection of independent modules, each of which is responsible for a separate material property (such as stiffness, dilatancy, etc.). Instead, all material constants are interwoven in the several simulations.

11.6 Response Envelopes

The mechanical response according to a particular constitutive equation can be visualised by means of the so-called response envelopes [33]. To obtain two-dimensional (2D) graphical representations, the response envelopes are restricted to axisymmetric stresses and stretchings. We furthermore restrict our consideration to rectilinear extensions so that we only consider principal stresses. Referring to a given stress $\mathbf{T}$, all axisymmetric unit stretchings are given by

$$\mathbf{D} = \begin{pmatrix} \sin\alpha & 0 & 0 \\ 0 & (\cos\alpha)/\sqrt{2} & 0 \\ 0 & 0 & (\cos\alpha)/\sqrt{2} \end{pmatrix}, \quad 0 \le \alpha < 360°, \quad |\mathbf{D}| = 1. \qquad (11.1)$$

Applying these stretchings to a material under stress $\mathbf{T}$ causes (with a particular constitutive equation) stress responses $\dot{\mathbf{T}}$. In a polar representation, all these stress responses define the response envelope. This curve is an informative representation of the response according to the considered constitutive equation.

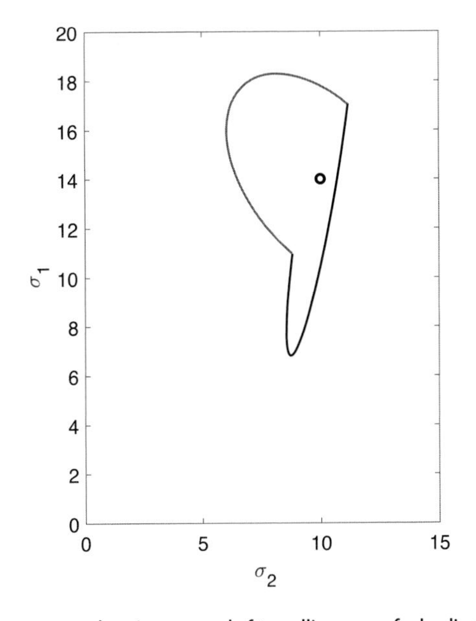

Figure 11.1 In plasticity theory the response envelope is composed of two ellipses, one for loading and one for unloading. The actual stress is located at the centre, which is common for both ellipses. The consistency condition guarantees that there is no gap between the two ellipses.

A constitutive law is expressed as $\dot{\mathbf{T}} = \mathbf{HD}$, where $\mathbf{H}$ can depend on $\mathbf{D}$, $\mathbf{H} = \mathbf{H}(\mathbf{D})$. As we consider unit $\mathbf{D}$ tensors, i.e. $\operatorname{tr} \mathbf{D}^2 = \operatorname{tr} \mathbf{D}^T\mathbf{D} = 1$, we obtain with $\mathbf{D} = \mathbf{H}^{-1}\dot{\mathbf{T}}$:

$$\operatorname{tr}(\mathbf{D}^T\mathbf{D}) = \operatorname{tr}[(\mathbf{H}^{-1}\dot{\mathbf{T}})^T\mathbf{H}^{-1}\dot{\mathbf{T}}] = \operatorname{tr}[\dot{\mathbf{T}}^T\mathbf{H}^{-T}\mathbf{H}^{-1}\dot{\mathbf{T}}] = 1. \tag{11.2}$$

This equation describes the response envelope. For linear (i.e. hypoelastic) constitutive laws, the matrix $\mathbf{B} := \mathbf{H}^{-T}\mathbf{H}^{-1}$ is constant and the equation $\operatorname{tr}(\dot{\mathbf{T}}^T\mathbf{B}\dot{)} = 1$ describes an ellipsoid in the space of principal stress rates. As stated in Chapter 14, elastoplastic constitutive equations are piecemeal linear, they consist of two linear relations, one for loading and one for unloading. Consequently, the pertinent response envelopes are composed of two ellipses, as shown in Fig. 11.1. At the limit state, the ellipse for loading degenerates to a straight line.

Note that the geometric representation by means of the response envelope is only affected by the symmetric part of the matrix $\mathbf{H}$.

11.7 Proportional Strain Paths

A convenient method to get an idea on the performance of a constitutive equation is to plot the stress paths predicted by this equation for proportional strain paths, i.e. strain paths with $\mathbf{D} = \text{const}$ For axisymmetric stress and strains the obtained stress paths can be plotted in a 2D stress space. The stretchings can be obtained from Equation 11.1 for, say, $\alpha = 0, 10°, 20°, \ldots$. If such stretchings start at $\mathbf{T} \neq \mathbf{0}$, they result in curved stress paths. For sand, no tensile stresses can ever be obtained, which

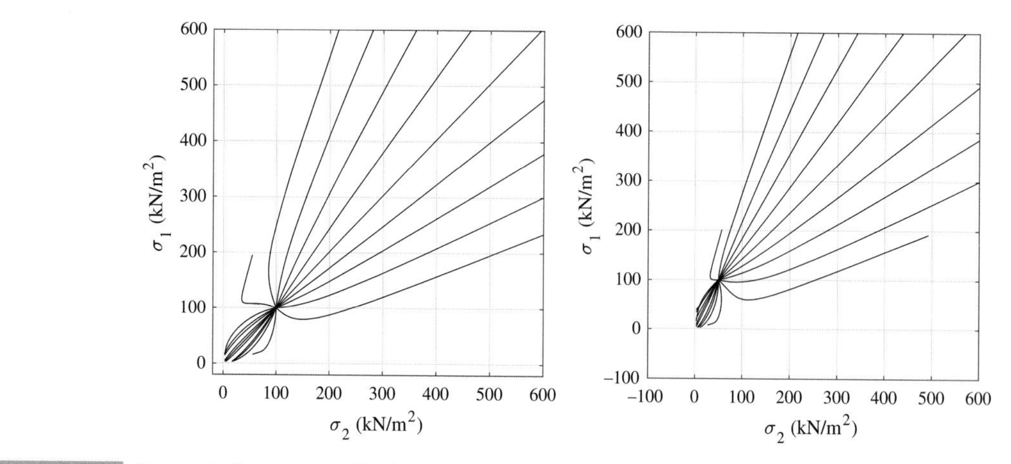

Figure 11.2 Stress paths for proportional loadings with $\mathbf{D} = \text{const}$ using the barodetic constitutive Equation 16.6.

implies that the stress paths must remain within the quadrant of compressive stresses, see Fig. 11.2.

11.8 Large Deformations

The consideration of non-linear terms in so-called theories of large deformation leads to mathematically complex expressions. Many problems in geotechnics deal with large deformations (landslides, penetration of probes, piles, etc.). It is then possible to use incremental constitutive equations without resorting to such theories, if the deformation is applied in sufficiently small steps and the reference configuration is updated accordingly.

Hypoplastic and barodetic constitutive equations are relations between the stress rate $\dot{\sigma}_{ij}$ and the deformation rate $\dot{\varepsilon}_{kl}$. Instead of $\dot{\varepsilon}_{kl}$ the stretching D_{kl} is used, which is derived from the velocity field v_i through $D_{ij} = \frac{1}{2}(v_{i,j} + v_{j,i})$ and, strictly speaking, cannot be considered as a time derivative of any deformation tensor (with the exception of logarithmic strain), which is only applicable to rectilinear extensions (i.e. for motions without principal axis rotation). Consequently, the relation $D_{ij} = \dot{\varepsilon}_{ij}$ is valid only approximately.

11.9 Role of Thermodynamics

Thermodynamics is an abstract part of physics, to which some people attribute almost magical properties, e.g. they try to derive constitutive equations from thermodynamics or, at least, they try to formulate restrictions upon such equations, so as to not violate thermodynamics.

However, constitutive equations cannot be derived from thermodynamics, because they express the specialities, which distinguish a substance (such as rubber) from another one (such as sand), whereas thermodynamics are valid for every material.

Classical thermodynamics is related to the consideration of a special constitutive law, namely the relation between pressure and volume, $p(V)$, for ideal gases and the observation that the compressibility of the gas depends on whether the compression is adiabatic or isothermal.

Essential statements of thermodynamics are the first law (energy conservation) and the second law (entropy production), of which many (and not necessarily congruent) versions exist.

In connection with constitutive equations for other solids (e.g. soil) the question arises on the extent to which restrictions can be derived from thermodynamics, i.e. whether they do not violate the rules of thermodynamics. It proves that thermodynamic considerations are not very helpful, because they refer to *cyclic* processes, where the constitutive equations proposed so far fail anyway. At best, it is possible to construct thermodynamic potentials in such a way as to derive existing constitutive equations.

11.10 Comparison of Constitutive Equations

In view of the many constitutive models proposed so far, researchers ask which one is the best. At first glance the validation of a constitutive model appears to be an easy task. As soon as a constitutive model is put forward, one should try to check whether this model can describe as many experimental results as possible. This means that these results have to be numerically simulated by means of the considered model. The closer the simulations ('predictions') are to the measured curves, the better is the constitutive model. A thorough examination proves, however, that this task is unfeasible. As known, there is no unique norm to validate (measure) the deviation of a curve $y(x)$ from another curve $y_1(x)$. The related ambiguity is increased if one has to consider many curves, i.e. the outcomes of different experiments. At that, one should take also into account the varying complexity of constitutive equations, which cannot be measured. There is no unique way to measure neither complexity nor simplicity and, moreover, there is no way to measure simplicity against quality of predictions. This is why several competitions of constitutive models remained non-conclusive [34, 90, 120].

12 Elasticity

12.1 Definition of Elasticity

Elasticity requires that the actual stress is a *function* of the actual deformation, i.e. it depends only on the actual deformation. This implies that the actual stress does *not* depend on the deformation history. Thus, any sort of path dependence and irreversible deformation are excluded.

12.2 Linear Elasticity

Linearity is not required by the definition stated above but implies a considerable simplification of elasticity. Linear elasticity is the simplest constitutive equation for solids. As a special case, the constitutive equation of Hooke applies to isotropic, linear-elastic materials (such as steel).

Remark Hooke expressed his law in latin: *ut tensio sic vis*, and used the anagram *ceiiinosssttuv* to hide it from competitors.

A linear dependence of σ_{ij} on ε_{ij} can be expressed as $\sigma_{ij} = E_{ijkl}\varepsilon_{kl}$. E_{ijkl} has $3^4 = 81$ components. Symmetry of σ_{ij} and ε_{kl} implies that there are only $6 \times 6 = 36$ independent components (material constants). If the elastic material is also *isotropic*, then there are only two independent material constants, λ and μ:

$$\sigma_{ij} = \lambda \varepsilon_{kk} \delta_{ij} + 2\mu \varepsilon_{ij} \tag{12.1}$$

with $\varepsilon_{kk} := \varepsilon_{11} + \varepsilon_{22} + \varepsilon_{33}$. λ and μ are the so-called Lamé constants. The inverse relation reads:

$$\varepsilon_{ij} = -\frac{\lambda}{2\mu} \frac{\sigma_{kk}}{(3\lambda + 2\mu)} \delta_{ij} + \frac{1}{2\mu} \sigma_{ij}. \tag{12.2}$$

There are many different representations of Hooke's law. With the shear modulus $G \equiv \mu$ or Young's modulus E and Poisson ratio ν we have:

$$\sigma_{ij} = 2G \left(\varepsilon_{ij} + \frac{\nu}{1 - 2\nu} \varepsilon_{kk} \delta_{ij} \right), \tag{12.3}$$

$$\sigma_{ij} = \frac{E}{1 + \nu} \varepsilon_{ij} + \frac{\nu E}{(1 + \nu)(1 - 2\nu)} \varepsilon_{kk} \delta_{ij}. \tag{12.4}$$

The inverse relations read:

$$\varepsilon_{ij} = \frac{1}{2G} \left(\sigma_{ij} - \frac{\nu}{1 + \nu} \delta_{ij} \sigma_{kk} \right) \quad \text{or} \quad \varepsilon_{ij} = \frac{1}{E} [(1 + \nu)\sigma_{ij} - \nu \delta_{ij} \sigma_{kk}]. \tag{12.5}$$

Hooke's law can be broken down into hydrostatic and deviatoric parts:

$$\text{hydrostatic:} \quad \sigma_{kk} = 3B\varepsilon_{kk}, \quad \text{deviatoric:} \quad \sigma_{ij}^* = 2\mu\varepsilon_{ij}^* \tag{12.6}$$

with the compression modulus $B := \lambda + \frac{2}{3}\mu$. Often, the symbol K is used instead of B.

The various material constants are interrelated, e.g.:

$$\nu = \frac{\lambda}{2(\lambda+\mu)}, \quad \lambda = \frac{\nu E}{(1+\nu)(1-2\nu)},$$

$$E = \frac{\mu(2\mu+3\lambda)}{\lambda+\mu}, \quad \mu \equiv G = \frac{E}{2(1+\nu)}, \quad B \equiv K = \frac{E}{3(1-2\nu)}.$$

Exercise 12.2.1 Derive Equation 12.2 from Equation 12.1.

Exercise 12.2.2 Which is the value of ν for incompressible materials?

Exercise 12.2.3 A linear-elastic rod is compressed by the strain ε_z in axial direction z. What is the strain ε_r in radial direction r? What is the stiffness $d\sigma_z/d\varepsilon_z$? Calculate the stiffness for the case where the lateral expansion/contraction is inhibited.

12.2.1 Derivation of Hooke's Law

Interestingly, Hooke's law can be derived from the definition of elasticity stated above and from linearity and isotropy using the representation theorem of Cayley–Hamilton for tensor-valued functions of one tensor-valued symmetric argument. $\mathbf{T}$ is a *function* of the deformation $\mathbf{F}$: $\mathbf{T} = \mathbf{f}(\mathbf{F})$. Isotropy requires that previous rotations remain undetectable, and this leads to

$$\mathbf{T} = \mathbf{f}(\mathbf{FQ}) = \mathbf{f}(\mathbf{F}). \tag{12.7}$$

With $\mathbf{F} = \mathbf{VR}$ and $\mathbf{Q} = \mathbf{R}^T$ we have:

$$\mathbf{T} = \mathbf{f}(\mathbf{VRQ}) = \mathbf{f}(\mathbf{V}).$$

Objectivity (see Section 18.2) requires that $\mathbf{f}$ be an isotropic function, i.e. $\mathbf{f}(\mathbf{QVQ}^T) = \mathbf{Qf}(\mathbf{V})\mathbf{Q}^T$. For such functions, the theorem of Cayley–Hamilton can be applied:

$$\mathbf{T} = \mathbf{f}(\mathbf{V}) = \alpha_0\mathbf{1} + \alpha_1\mathbf{V} + \alpha_2\mathbf{V}^2,$$

where α_i are scalar-valued functions (so-called invariants) of $\mathbf{V}$. With the infinitesimal strain $\mathbf{E}$ and geometric linearisation it follows: $\mathbf{V} = \mathbf{1} + \mathbf{E}$, $\mathbf{V}^2 \approx \mathbf{1} + 2\mathbf{E}$, hence

$$\mathbf{T} = \mathbf{g}(\mathbf{E}) = \beta_0\mathbf{1} + \beta_1\mathbf{E}.$$

Rendering $\mathbf{g}$ linear in $\mathbf{E}$ leads finally to

$$\mathbf{T} = \lambda(\text{tr}\mathbf{E})\mathbf{1} + 2\mu\mathbf{E}.$$

12.3 Modifications of Elasticity

To take into account that the stiffness of soil increases with stress level, Young's (or the shear) modulus is often modified as, e.g. $E = E_0(\sigma/p_0)^\beta$ with $\sigma := \sqrt{\sigma_{ij}\sigma_{ij}}$.

However, a relation $\mathbf{T} = \mathbf{f}_1(\mathbf{T}, \mathbf{E})$ cannot always be transformed into $\mathbf{T} = \mathbf{f}(\mathbf{E})$, and thus, it is not necessarily elastic in the sense stated above. This becomes obvious in cyclic processes, where stress or strain cycles do not lead to the initial strain or stress, respectively. Consider, for example, the stress cycle: (i) hydrostatic stress Δp, (ii) deviatoric stress $\Delta\sigma^*_{ij}$, (iii) hydrostatic stress $-\Delta p$ and (iv) deviatoric stress $-\Delta\sigma^*_{ij}$. The deviatoric strain of step (ii) overshoots the one obtained with step (iv), because the stiffness at (iv) is lower than at step (ii).

Deriving Equation 12.8 with respect to time leads erroneously to $\dot{\mathbf{T}} = \lambda(\mathrm{tr}\mathbf{D})\mathbf{1} + 2\mu\mathbf{D}$. This equation is improper, because (i) $\mathbf{D} \neq \dot{\mathbf{E}}$ and (ii) $\dot{\mathbf{T}}$ is not an objective quantity (see Section 18.2.4).

Truesdell introduced hypo-elasticity in 1953 [109] as a relation $\overset{\circ}{\mathbf{T}} = \mathbf{h}(\mathbf{T}, \mathbf{D})$ that is *linear* in $\mathbf{D}$. Linearity excludes irreversible behaviour, as it implies equal stiffness at loading and unloading: $\mathbf{h}(\mathbf{T}, -\mathbf{D}) = -\mathbf{h}(\mathbf{T}, \mathbf{D})$. Interestingly, a yield stress, i.e. a state with $\dot{\mathbf{T}} = \mathbf{0}$ for particular stretchings, could be described. Truesdell concluded:

> ... failure or yield-like phenomenon is in principle predicted, within the framework of hypo-elasticity, by quantities measurable in experiments on small strain. This feature is to be contrasted with the usual theories of plasticity, in which the yield stress is of a magnitude assumed ‘a priori’ and altogether inaccessible to experiments at smaller stresses.
>
> The theory of hypo-elasticity has never been advocated as a theory for describing any particular material ... we neither claim to explain any particular experiment nor imply any connection with the usual theories of plasticity.

12.4 Elasticity in Soil Mechanics

In the early days of soil mechanics, linear elasticity was considered as *the* constitutive equation for solids. Further, the simplicity of Hooke’s law allows the (still difficult) analytical solution of some boundary value problems [68, 82, 84, 86, 112], such as

- Boussinesq: vertical force F on the surface of an elastic halfspace
- Cerrutti: tangential force on the surface of the halfspace
- Mindlin: force inside the halfspace
- Lamé: thick-walled tube under internal and external pressures
- Hertzian pressure: contact of two elastic bodies (Fig. 12.1).

Elastic solutions are useful as reference solutions and are used in soil mechanics, with a limited range of applicability. At least they yield stress fields that fulfil the boundary conditions and the equations of equilibrium. In particular, the Boussinesq’s solution is used to calculate the stress field underneath foundations. According to this solution, the vertical stress at depth z below F reads $\sigma_z = 3F/(2\pi z^2)$ and is thus independent of the elastic constants E and ν. Therefore, it is believed to be a good estimation also for inelastic materials, such as soil, and is used to calculate the settlement of foundations. To this end, $\sigma_z(z)$ is calculated with Boussinesq’s elastic solution, but the corresponding strain $\varepsilon_z(\sigma_z)$ is estimated on the basis of oedometric test results with the considered soil. Note that the assumption of a linear elastic halfspace yields an infinite settlement of the load F.

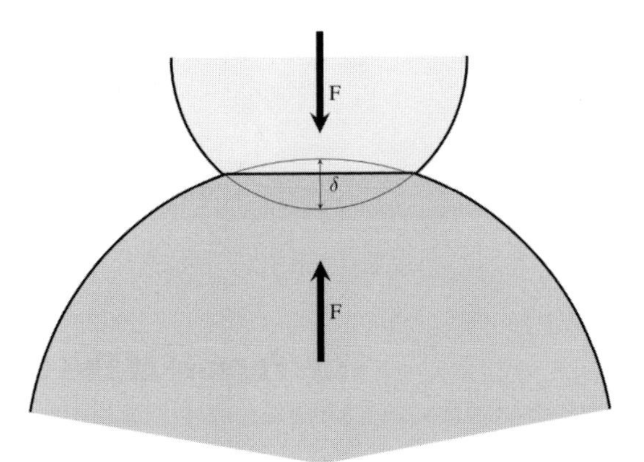

Figure 12.1 Hertzian pressure [112]: two elastic bodies with curved and frictionless surfaces are pressured against each other. The contact area as well as the indentation δ increase with the pressure force $\mathbf{F}$.

Later, more attention was paid to the fact that soil undergoes irreversible deformation, i.e. the deformation depends not only on the actual stress but also on the stress history, and this effect is not covered by the theory of elasticity, neither are limit states (i.e. vanishing stiffness) nor dilatancy/contractancy covered. However, linear elasticity is the basis for calculations of wave propagation in soil dynamics, see Chapter 13. The justification for this is the fact that cyclic deformation of small amplitude becomes elastic after a few cycles.

Elasticity is a basic ingredient of elastoplastic constitutive equations. They are based on the idea that deformations are initially elastic, whereas irreversible deformation sets in gradually with loading, and unloading is purely elastic.

13 Elastic Waves

13.1 Purpose of This Chapter

Waves constitute a huge field of mathematical physics. Here only a few aspects will be presented, as they pertain to soil mechanics in the following questions: What does the term 'elastic waves' mean and how do they propagate in soil strata? How does propagate a mechanical action (e.g. an impact) along a pile? From a theoretical point of view, waves are a consequence of the conservation laws.

13.2 What Are Waves?

Although waves have great importance in almost all areas of science and technology, a general definition of waves is still missing [117].

An action (e.g. a displacement) upon a particular point at the boundary of a body is also noticeable in distant places. Often, the propagation of this action is neglected, and the deformation of remote parts is considered as simultaneous with the action. This is the case when we consider slow (so-called quasi-static) loading. Otherwise, we have to consider loading as a process in time, i.e. as a so-called initial boundary value problem (IBVP), and to take inertia into account. Every load or displacement at the boundary of a body is communicated to the inside of the body via waves. In the quasi-static processes, waves are disregarded, and it is assumed that the final state, which sets in after waves and vibrations are damped, occurs simultaneously with the action (i.e. the acceleration term in Equation 8.8 is neglected). If the imposed displacement is faster than the speed of the wave, the stiffness of the body does not come into play and the body only opposes the impacting mass with its inertia and thus behaves like an ideal fluid (so-called transonic impact).

The solution of an IBVP describes the propagation of a spatially limited action upon a body, and this is called a wave: a configuration (or a signal) that travels in a continuous medium. It can undergo changes but must remain recognisable as such. Consider, for example, tectonic ruptures in the Earth's crust, the accompanying waves travel through rock and soil strata and can be amplified resulting in catastrophic earthquakes. Waves are reflected at the boundaries and interact with the new incoming waves, if the excitation is continued in time. This results in so-called standing waves, which is a synonym of oscillation of the considered body.

Mechanical waves are closely linked to the conservation laws of mass and momentum. However, there are also other types of waves. The so-called kinematic waves result from mass conservation only. Moving discontinuity surfaces (see Section 5.4) are also considered as waves.

13.3 Kinematic Waves

The following differential equation describes one-dimensional (1D) convection:

$$u_t = -v u_x, \quad -\infty < x < \infty, \quad 0 < t < \infty, \tag{13.1}$$

where u is the concentration of particles entrained in a flow of velocity v (diffusion is not taken into account). Example: At time $t = 0$ we pass a (non-diffusive) contaminant with concentration $u(x,0) = \varphi(x)$ into a river. We are looking for the concentration distribution $u(x,t)$ at later times ($t > 0$). The solution $u(x,t) = \varphi(x - vt)$ is called a kinematic wave and satisfies the differential equation (Equation 13.1).

Conservation equations have the form

$$u_t + f_x = 0. \tag{13.2}$$

Here, $u(x,t)$ is a field quantity (usually a density), and f is its flow rate ('flux'). If we express the propagation velocity by v, then $f = u \cdot v$. If we consider, for example, road traffic, then $u(x,t)$ is the density of cars at location x and $f(x,t)$ is the number of cars passing the location x in the unit of time. Strictly speaking, the distribution of cars on a road cannot be described by a continuous function. The use of continuous functions facilitates the analysis of the processes. It is even exact when we refer not to the number of cars, but to its expected value.

The differential equation (Equation 13.2) can be interpreted as a conservation law: we require that the volume integral $\int_B u \, \mathrm{d}V$ remains unchanged in time. B is a material domain, whose boundaries are variable in time. If, for example, ρ is the density of a substance, then $\int_B \rho \, \mathrm{d}V$ is the mass of a body, which is to remain constant in time. The latter is expressed by the requirement that the so-called material time derivative of $\int_B u \, \mathrm{d}V$ should vanish:

$$\frac{\mathrm{d}}{\mathrm{d}t} \int_B u \, \mathrm{d}V = \int_B \left(\frac{\mathrm{d}u}{\mathrm{d}t} + u \cdot \operatorname{div} \mathbf{v} \right) \mathrm{d}V = 0. \tag{13.3}$$

The above equation follows from Reynolds transport theorem ($\mathbf{v}$ is the velocity field). Since it is supposed to hold for arbitrary domains B, the integrand itself must vanish (provided it is continuous):

$$\frac{\mathrm{d}u}{\mathrm{d}t} + u \cdot \operatorname{div} \mathbf{v} = 0. \tag{13.4}$$

With $\nabla u \cdot \mathbf{v} + u \operatorname{div} \mathbf{v} = \operatorname{div}(u\mathbf{v})$ and with $\mathrm{d}u/\mathrm{d}t := \partial u/\partial t + (\partial u/\partial x) \cdot v$ it follows:

$$\frac{\partial u}{\partial t} + \operatorname{div}(u\mathbf{v}) = 0. \tag{13.5}$$

If only one dimension is considered ($\mathbf{v} \equiv v$), it follows with $f = uv$ the differential equation (Equation 13.2).

13.4 Elastic Waves in One-dimensional Continua

As shown in Chapter 8, the balance of mass and momentum leads to the equation

$$\rho\left(\frac{\partial \mathbf{v}}{\partial t} + \mathbf{v}\cdot\nabla\mathbf{v}\right) - \operatorname{div}\mathbf{T} = \rho\mathbf{b}, \tag{13.6}$$

and in one spatial dimension:

$$\rho\left(\frac{\partial v}{\partial t} + v\cdot\frac{\partial v}{\partial x}\right) - \frac{\partial \sigma}{\partial x} = \rho b. \tag{13.7}$$

For low values of the velocity v, the non-linear term $v\cdot\frac{\partial v}{\partial x}$, i.e. the convective acceleration, can be neglected. To show this, we consider a harmonic wave $v = v_0\cos(kx-\omega t)$ and realise that the amplitude of $v\frac{\partial v}{\partial x}$ in Equation 13.7 has the value kv_0^2, while the amplitude of $\frac{\partial v}{\partial t}$ has the value ωv_0. Thus, the convective acceleration can be neglected if the ratio $M := kv_0^2/(\omega v_0) = v_0/c$ is small, where $c = \omega/k$ is the wave velocity, also called phase velocity. M is the Mach number [9]. Gas dynamics investigates phenomena in which the convective term $v\frac{\partial v}{\partial x}$ cannot be neglected.

We now consider the compression of an elastic rod. From $\varrho\dfrac{\partial v}{\partial t} = \dfrac{\partial \sigma}{\partial x} + b$ we obtain with $\sigma = E\varepsilon$ and $c := \sqrt{E/\varrho}$:

$$\frac{\partial v}{\partial t} - c^2\frac{\partial \varepsilon}{\partial x} = \frac{b}{\rho}, \tag{13.8}$$

where σ is the stress acting in a bar (here σ and ε are positive at tension) and E is Young's modulus of the bar. Insertion of $\varepsilon = \partial u/\partial x$ and $v = \partial u/\partial t$ into (Equation 13.8) yields

$$\frac{\partial^2 u}{\partial t^2} - c^2\frac{\partial^2 u}{\partial x^2} = \frac{b}{\rho}. \tag{13.9}$$

One can eliminate a time-independent mass force b by the variable substitution $u := u + \bar{u}$, where $\bar{u}$ is independent of t and satisfies the quasi-static problem $E\dfrac{\partial^2 \bar{u}}{\partial x^2} = b$. This yields the so-called wave differential equation

$$\frac{\partial^2 u}{\partial t^2} - c^2 \frac{\partial^2 u}{\partial x^2} = 0. \tag{13.10}$$

Remark 1 Regarding pulses in elastic rods, Equation 13.10 is only applicable if the external disturbance producing the pulse is applied relatively slowly [2].

Remark 2 If one considers bending waves in a beam, the differential equation is more complicated: $\varphi_{tt} + \gamma^2 \varphi_{xxxx} = 0$. Bending waves exhibit dispersion, i.e. $\omega = \pm\gamma k^2$.

The following solutions apply to the wave differential equation:

1. General solution (Laplace):

$$u(x,t) = \varphi(x - ct) + \psi(x + ct),$$

where φ and ψ are arbitrary functions to be determined from the initial and boundary conditions. $\varphi(x)$ and $\psi(x)$ are displacement distributions which move with velocity c in the direction of increasing and decreasing x, respectively.
2. Solution with the method of *characteristics:*
Neglecting small terms, mass conservation (Equation 8.2) can be expressed as $\partial\varepsilon/\partial t - \partial v/\partial x = 0$ (i). Equally, momentum conservation (Equation 8.3) can be expressed as $c\partial(c\varepsilon)/\partial x - \partial v/\partial t = 0$ (ii). These are two coupled equations, but the equations c(i)–(ii) and c(i)+(ii):

$$\frac{\partial(v + c\varepsilon)}{\partial t} - c\frac{\partial(v + c\varepsilon)}{\partial x} = 0, \tag{13.11}$$

$$\frac{\partial(v - c\varepsilon)}{\partial t} + c\frac{\partial(v - c\varepsilon)}{\partial x} = 0, \tag{13.12}$$

are uncoupled, with the variable $(v + c\varepsilon)$ in Equation 13.11 only and the variable $(v - c\varepsilon)$ in Equation 13.12 only. Now we introduce in the x-t-plane the parallel lines $x+ct = s_1 =$ const and the lines $x-ct = s_2 =$ const Each line is called an s_1 or s_2-characteristic, respectively. Equations 13.11 and 13.12 can now be written as:

$$\frac{\mathrm{d}(v + c\varepsilon)}{\mathrm{d}s_1} = 0, \tag{13.13}$$

$$\frac{\mathrm{d}(v - c\varepsilon)}{\mathrm{d}s_2} = 0. \tag{13.14}$$

This means that the quantity $v + c\varepsilon$ remains constant along an s_1-characteristic, and equally the quantity $v - c\varepsilon$ remains constant along an s_2-characteristic. We also say that the quantities $v + c\varepsilon$ and $v - c\varepsilon$ propagate along the characteristics.
Thus, we can prescribe along an infinitely long rod ($-\infty < x < \infty$) the initial distributions $v(x, t = 0)$ and $\varepsilon(x, t = 0)$. Using Equations 13.13 and 13.14 we can then easily obtain for any time $t > 0$ the v- and ε-values at any location x.
3. Solution by d'Alembert: For an infinitely long rod, the initial distributions of the displacement $u(x, 0) = f(x)$ and velocity $v(x, 0) = g(x)$ may be given. This is an initial value problem with the solution

$$u(x,t) = \frac{1}{2}\left[f(x - ct) + f(x + ct)\right] + \frac{1}{2c}\int_{x-ct}^{x+ct} g(z)\,\mathrm{d}z. \tag{13.15}$$

4. Excitation at the end of a semi-infinite rod: For the so-called semi-infinite rod ($0 \leq x < \infty$), a conceivable boundary value problem is to prescribe the displacement at the boundary $x = 0$ as a function of t:

$$u(0, t) = f(t), \qquad t \geq 0.$$

The solution is simply that this displacement propagates with velocity c in the direction of increasing x:

$$u(x, t) = \begin{cases} f(t - x/c) & \text{for} \quad 0 < x \leq ct, \\ 0 & \text{for} \quad x > ct. \end{cases}$$

13.4.1 Transmission and Reflection

The effect of a sudden change in the properties of a rod (cross section A and Young's modulus E) will now be investigated. Starting from the general solution for the displacement field $u = \varphi(x - ct) + \psi(x + ct)$ we get

$$\varepsilon = \frac{\partial u}{\partial x} = \varphi'(x - ct) + \psi'(x + ct) \tag{13.16}$$

and

$$v = \frac{\partial u}{\partial t} = -c\varphi'(x - ct) + c\psi'(x + ct). \tag{13.17}$$

We now introduce the following quantities

$$\varepsilon_\varphi := \varphi'(x - ct), \quad v_\varphi := -c\varphi'(x - ct) = -c\varepsilon_\varphi, \tag{13.18}$$

$$\varepsilon_\psi := \psi'(x + ct), \quad v_\psi := c\psi'(x + ct) \quad = c\varepsilon_\psi, \tag{13.19}$$

and obtain

$$\varepsilon = \varepsilon_\varphi + \varepsilon_\psi, \quad v = v_\varphi + v_\psi = -c(\varepsilon_\varphi - \varepsilon_\psi). \tag{13.20}$$

The internal force F (positive for compression) reads:

$$F := -A\sigma = -AE\varepsilon = F_\varphi + F_\psi \tag{13.21}$$

with $F_\varphi = -AE\varepsilon_\varphi = \frac{AE}{c}v_\varphi$ and $F_\psi = -AE\varepsilon_\psi = -\frac{AE}{c}v_\psi$. The quotient AE/c is called impedance Z, $Z = \frac{AE}{c} = Ac\varrho$. One can therefore write: $F_\varphi = Zv_\varphi$, $F_\psi = -Zv_\psi$ i.e.:

$$v = v_\varphi + v_\psi = \frac{1}{Z}\left(F_\varphi - F_\psi\right). \tag{13.22}$$

Let the impedance change abruptly at the point $x = x_a$:

$$\text{range 1:} \qquad x < x_a, \quad Z = Z_1, \quad F = F_1, \quad v = v_1 \tag{13.23}$$

$$\text{range 2:} \qquad x > x_a, \quad Z = Z_2, \quad F = F_2, \quad v = v_2. \tag{13.24}$$

Continuity at x_a requires Equations 13.21 and 13.22 as follows:

$$F_{\varphi 1} + F_{\psi 1} = F_{\varphi 2} + F_{\psi 2}, \tag{13.25}$$

$$\frac{1}{Z_1}\left(F_{\varphi 1} - F_{\psi 1}\right) = \frac{1}{Z_2}\left(F_{\varphi 2} - F_{\psi 2}\right). \tag{13.26}$$

These equations link the forces running towards the discontinuity, $F_{\varphi 1}$ und $F_{\psi 2}$, with those ones running away from it, $F_{\varphi 2}$ und $F_{\psi 1}$. Equations 13.25 and 13.26 can also be formulated as follows:

$$F_{\psi 1} = a_{11}F_{\varphi 1} + a_{12}F_{\psi 2}, \quad (13.27)$$

$$F_{\varphi 2} = a_{21}F_{\varphi 1} + a_{22}F_{\psi 2}, \quad (13.28)$$

with the *force reflection coefficients*

$$a_{11} = \frac{Z_2 - Z_1}{Z_2 + Z_1}, \quad a_{22} = \frac{Z_1 - Z_2}{Z_1 + Z_2} \quad (13.29)$$

and the *force transmission coefficients*

$$a_{12} = \frac{2Z_1}{Z_2 + Z_1}, \quad a_{21} = \frac{2Z_2}{Z_1 + Z_2}. \quad (13.30)$$

If one expresses Equations 13.25 and 13.26 with the aid of the velocities one obtains the following relations:

$$Z_1 v_{\varphi 1} - Z_1 v_{\psi 1} = Z_2 v_{\varphi 2} - Z_2 v_{\psi 2}, \quad (13.31)$$

$$v_{\varphi 1} + v_{\psi 1} = v_{\varphi 2} + v_{\psi 2}, \quad (13.32)$$

or

$$v_{\psi 1} = b_{11}v_{\varphi 1} + b_{12}v_{\psi 2}, \quad (13.33)$$

$$v_{\varphi 2} = b_{21}v_{\varphi 1} + b_{22}v_{\psi 2}, \quad (13.34)$$

with the *velocity reflection coefficients*

$$b_{11} = \frac{Z_1 - Z_2}{Z_1 + Z_2}, \quad b_{22} = \frac{Z_2 - Z_1}{Z_1 + Z_2} \quad (13.35)$$

and the *velocity transmission coefficients*

$$b_{12} = \frac{2Z_2}{Z_2 + Z_1} = a_{21}, \quad b_{21} = \frac{2Z_1}{Z_1 + Z_2} = a_{12}. \quad (13.36)$$

The following limit cases can be considered:
Free end: $Z_2 = 0 \rightsquigarrow a_{11} = -1,\ b_{11} = 1$. Herewith, one obtains: $F_{\psi 1} = -F_{\varphi 1}, v_{\psi 1} = v_{\varphi 1}$, i.e. the wave reflected at the free end of the rod carries the same velocity but the opposite force (or strain).
Fixed end: $Z_2 = \infty \rightsquigarrow a_{11} = 1,\ b_{11} = -1$, from which follows: $F_{\psi 1} = F_{\varphi 1}, v_{\psi 1} = -v_{\varphi 1}$, i.e. the force (or strain) returns with the same sign, while the velocity suffers a sign reversal.

In oscillation problems, the trigonometric functions with their troublesome addition theorems can be avoided by using complex numbers. The basis for this is Euler's equation:

$$e^{i\varphi} = \cos\varphi + i\sin\varphi. \quad (13.37)$$

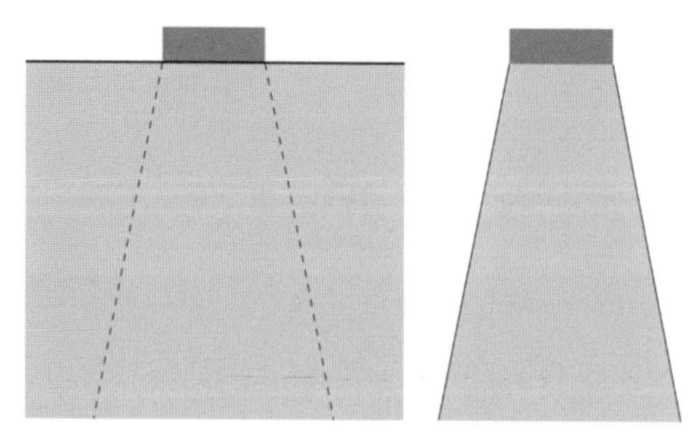

Figure 13.1 A rigid block rests upon the elastic halfspace, the relevant part of which can be represented approximately by a cone-like section. This cone can be conceived as a rod with variable cross section. In this sense, the response of the halfspace upon the block can be represented by a spring and a dashpot [118].

13.4.2 Dynamic Stiffness of a Rod

Let us consider a semi-infinite rod, at the end $x = 0$ on which acts the force $F(t) = F_0 \exp(i\omega t)$. We are looking for the relationship between the force $F(t)$ and the displacement $u_0(t)$ at $x = 0$, where the ratio F/u is called stiffness. The solution of the wave equation is $u(x,t) = \varphi(x - ct)$. The second term $\psi(x + ct)$ does not play any role here. The solution takes the form $u = u_0 \exp[-ik(x - ct)]$ with $k := \omega/c$. With $F(t) = AE\varepsilon(x = 0, t) = AE \left.\frac{\partial u}{\partial x}\right|_{x=0} = AEkiu(t)$ the complex stiffness K in this case is imaginary

$$K := \frac{F(t)}{u(t)} = iAEk = i\omega\rho cA, \tag{13.38}$$

which indicates a phase shift between $F(t)$ and $u(t)$ of 90°. The result is exactly the same as if the force $F(t)$ were acting on a viscous damper.

So, if we cut off a rod and replace the right part with a viscous damper of viscosity ρcA, then the left part will not 'notice' any change (i.e. no reflection, full transmission). This 1D case can be generalised for spatial interfaces. The medium on one side of the interface can be replaced by viscous dashpots with the viscosity ρc per unit area. The medium on the other side then 'notices' no change. Applications: shielding of waves, simulation of boundaries in finite elements.

For more complicated ratios (inhomogeneous semi-infinite rod or semi-infinite rod with variable cross section) K is a complex number:

$$K = \bar{K} + i\omega C. \tag{13.39}$$

The inhomogeneous rod is therefore dynamically equivalent to a spring (spring constant $\bar{K}$) and a dashpot of viscosity C (Fig. 13.1).

13.5 Waves in Bodies of Finite Dimensions

The wave differential equation characterises propagation processes which take place in infinitely extended bodies. In bodies of limited extensions, the waves are reflected at the boundaries and can interact with the new incoming waves. This process can lead to the amplification of their action (resonance). The application of the wave solution is then possible, but complicated. In the case of bodies with limited extensions, we often do not speak of waves, but of oscillations (or standing waves) of systems with an infinite number of degrees of freedom. The solutions of initial boundary value problems are based on the so-called separation method (separation of variables).

13.5.1 Rod with Harmonic Excitation

We consider a rod with cross section area A, density ρ, Young's modulus E, longitudinal coordinate x, $0 \leq x \leq l$, elongation u and harmonic excitation $u(0,t) = u_0 e^{i\Omega t}$ at $x=0$. We search the function $u(x,t)$, for which we assume the representation $u(x,t) = X(x) \cdot T(t)$. No force acts at the free end $x = l$. Thus, the boundary conditions are:

$$x = 0: \quad u(0,t) = u_0 e^{i\Omega t} \tag{13.40}$$

$$x = l: \quad \frac{\partial u}{\partial x} = 0. \tag{13.41}$$

Equation 13.40 leads to $X(0) \cdot T(t) = u_0 e^{i\Omega t}$, hence $T(t) = u_0 e^{i\Omega t}$ and $X(0) = 1$. Introducing $u(x,t) = X(x) \cdot T(t)$ into Equation 13.10 yields $X + \left(\frac{c}{\Omega}\right)^2 X'' = 0$ with the solution

$$X(x) = a\, e^{i\frac{\Omega}{c}x} + b\, e^{-i\frac{\Omega}{c}x}. \tag{13.42}$$

Introducing Equation 13.42 into the boundary conditions Equations 13.40 and 13.41 yields:

$$X(0) = a + b = 1, \tag{13.43}$$

$$X'(l) = \frac{i\Omega}{c}\left(ae^{i\frac{\Omega l}{c}} - be^{-i\frac{\Omega l}{c}}\right) = 0 \leadsto b = ae^{i\frac{2\Omega l}{c}} \leadsto a - b = a\left(1 - e^{i\frac{2\Omega l}{c}}\right). \tag{13.44}$$

With the abbreviation $\xi := \Omega l/c$ and using the equation of Euler it follows:

$$a = \frac{1}{1 + e^{2\xi i}} = \frac{e^{-\xi i}}{e^{\xi i} + e^{-\xi i}} = \frac{e^{-\xi i}}{2\cos\xi}, \tag{13.45}$$

$$b = 1 - a = \frac{e^{\xi i}}{e^{\xi i} + e^{-\xi i}} = \frac{e^{\xi i}}{2\cos\xi}. \tag{13.46}$$

Hence, the displacement at $x = l$ reads

$$u(l,t) = X(l) \cdot T(t) = \frac{1}{\cos\xi}\, u_0\, e^{i\Omega t}. \tag{13.47}$$

The factor $\mathcal{A} := 1/\cos\xi$ is the so-called dynamic amplification factor. Obviously, $\mathcal{A} \to \infty$ for $\xi \to \pi/2$. Thus, resonance ($\mathcal{A} \to \infty$) is obtained for

$$\Omega = \frac{\pi c}{2l}. \tag{13.48}$$

Comments:

1. The rod can be conceived as a thin vertical strip of an elastic soil layer resting upon a bedrock. The harmonic excitation can be considered as an incoming earthquake, i.e. a harmonic displacement of the bedrock. Clearly, this presupposes an infinite

(or very long) duration of the earthquake. Depending on Ω, E and l, a considerable amplification of the earthquake can be obtained.

2. If u denotes the *horizontal* displacement and E is replaced by G, the equations presented above apply to shear waves.
3. Considering torsion of the rod with u now being the torsion angle and applying the boundary conditions of fixed end at $x = 0$ and harmonic rotary excitation at $x = l$, we can describe the so-called *resonant column test*. From the first (low-shear) resonance frequency f_1, the shear modulus of the rod can be obtained (without derivation): $G = 16\varrho l^2 f_1^2$.

13.6 Body Waves

We consider elastic waves in a 3D space. Equation 13.6 with linearisation ($\mathbf{v}\cdot\nabla\mathbf{v} = \mathbf{0}$), absence of volume forces ($\mathbf{b} = \mathbf{0}$) and $\mathbf{v} = \dot{\mathbf{u}}$ ($\mathbf{u}$ = displacement) reads: $\rho\dot{\mathbf{v}} = \operatorname{div}\mathbf{T}$. With Hooke's law we now express the stress $\mathbf{T}$ via the displacement $\mathbf{u}$ and obtain:

$$\rho\ddot{\mathbf{u}} = \mu\nabla^2\mathbf{u} + (\lambda + \mu)\nabla(\nabla\cdot\mathbf{u}). \tag{13.49}$$

A vector field $\mathbf{u}$ can always be represented as the sum of a rotation-free field $\mathbf{u}_p$ ($\operatorname{rot}\mathbf{u}_p = \nabla\times\mathbf{u}_p = \mathbf{0}$) and a divergence-free field $\mathbf{u}_s$ ($\nabla\cdot\mathbf{u}_s = 0$). Noting that a rotation-free field can be expressed as $\mathbf{u}_p = \nabla\Psi$ we obtain from Equation 13.49:

$$\rho(\ddot{\mathbf{u}}_p + \ddot{\mathbf{u}}_s) = \mu\nabla\cdot\nabla(\nabla\Psi + \mathbf{u}_s) + (\lambda + \mu)\nabla(\nabla\cdot\nabla\Psi). \tag{13.50}$$

Using the identity $\partial_j\partial_j\partial_i\Psi = \partial_j\partial_i\partial_j\Psi$ we obtain $\nabla\cdot\nabla(\nabla\Psi) = \nabla(\nabla\cdot\nabla\Psi) = \nabla^2\mathbf{u}_s$. Herein, we have used the abbreviations $\partial_i = \partial/\partial x_i$ and $\nabla^2\mathbf{u}_s = \nabla\cdot\nabla\mathbf{u}_s$. Thus, we have:

$$\rho(\ddot{\mathbf{u}}_p + \ddot{\mathbf{u}}_s) = \mu\nabla^2\mathbf{u}_s + (\lambda + 2\mu)\nabla^2\mathbf{u}_p \tag{13.51}$$

which can be reduced to

$$\rho\ddot{\mathbf{u}}_p = (\lambda + 2\mu)\nabla^2\mathbf{u}_p \quad \text{and} \quad \rho\ddot{\mathbf{u}}_s = \mu\nabla^2\mathbf{u}_s, \tag{13.52}$$

the differential equations for longitudinal (P) and transversal or shear waves (S), respectively. Their propagation velocities are (see Table 13.1)

$$c_p = \sqrt{\frac{\lambda + 2\mu}{\rho}} \quad \text{and} \quad c_s = \sqrt{\frac{\mu}{\rho}}. \tag{13.53}$$

The solution for the P-wave is:

$$\mathbf{u} = \mathbf{u}_p = \begin{pmatrix} u_x \\ u_y \\ u_z \end{pmatrix} = A_p \begin{pmatrix} l_x \\ l_y \\ l_z \end{pmatrix} \exp\left[i\omega\left(t - \frac{s}{c_p}\right)\right] \tag{13.54}$$

with $s := \mathbf{l}\cdot\mathbf{x}$, where $\mathbf{x}$ is the location vector and $\mathbf{l}$ is a unit vector in the propagation direction of the wave.

The solution for the S-wave is

$$\mathbf{u} = \mathbf{u}_s = \begin{pmatrix} u_x \\ u_y \\ u_z \end{pmatrix} = \underbrace{\begin{pmatrix} m_z A_{sy} - m_y A_{sz} \\ m_x A_{sz} - m_z A_{sx} \\ m_y A_{sx} - m_x A_{sy} \end{pmatrix}}_{\mathbf{A}_s \times \mathbf{m}} \exp\left[i\omega\left(t - \frac{s}{c_s}\right)\right], \tag{13.55}$$

where $\mathbf{m}$ is a unit vector ($|\mathbf{m}| = 1$), which is perpendicular to the vector $\mathbf{A}_s$ ($\mathbf{m} \cdot \mathbf{A}_s = 0$). s is given by the equation $s := \mathbf{m} \cdot \mathbf{x} = m_x x + m_y y + m_z z$. Consequently, the movement is perpendicular to the direction of propagation $\mathbf{m}$. If it is horizontal or vertical, we have an SH- or SV-wave, respectively.

In full space there are only P- and S-waves. All other waves follow from special boundary conditions.

Table 13.1. Typical shear wave velocities

Material	c_s (m/s)
Soft clay, loose sand	≤150
Medium stiff clay	250
Stiff clay, dense sand	350
Hard clay, very dense sand	450
Soft rock	600
Weathered rock	1000
Rock	>1500

13.7 Rayleigh Waves

In full space, S- and P-waves propagate independently of each other. However, the absence of stress on the surface of a halfspace causes a coupling between the two types of waves. The result is a Rayleigh wave. We investigate so-called monochromatic planar ($u_y \equiv 0$) waves of the following type:

$$\mathbf{u} = \mathbf{u}_0\, e^{i(kx-\omega t)} f(z), \tag{13.56}$$

where z is the height coordinate and $|f(z)|$ decays with depth (i.e. for $z \to -\infty$). For $k^2 - \left(\frac{\omega}{c_{p/s}}\right)^2 < 0$ one would obtain solutions, which are periodic with respect to z.

After some calculation [64] one obtains with $\xi := \dfrac{\omega}{c_s k}$ the equation

$$\xi^6 - 8\xi^4 + 8\xi^2\left(3 - 2\frac{c_s^2}{c_p^2}\right) - 16\left(1 - \frac{c_s^2}{c_2^p}\right) = 0. \tag{13.57}$$

From Equation 13.56 one can see, that $c_R := \omega/k$ is the propagation speed of the Rayleigh wave. Thus, ξ is the ratio c_R/c_s. It can be seen from Equation 13.57 that ξ

depends only on the ratio c_s/c_p or on the Poisson number ν, since

$$\frac{c_s^2}{c_p^2} = \frac{1-2\nu}{2(1-\nu)}$$

holds. Equation 13.57 has only one real positive root, which can be represented as a function of ν. Its value specifies the propagation velocity c_R, and accordingly, the proportionality between wavenumber $k = 1/\lambda_R$ (λ_R is the wavelength) and frequency ω: the greater the frequency, the greater the wavenumber, and the smaller the wavelength. The dependence of the wavelength on the frequency is called dispersion. The amplitudes decay exponentially with depth. With increasing frequency, the 'penetration depth' of a Rayleigh wave decreases (the 'skin effect'). Rayleigh waves occur at a small depth from the surface (practically $z \leq 1.5\lambda_R$) and propagate along the surface with a velocity $c_R < c_s$. The ratio c_R/c_s depends on the Poisson number ν and since this varies between 0.25 and 0.5 for soils, one has on average $c_R \approx 0.94c_s$. There is a phase shift of 90° between horizontal and vertical displacement, so that the particle trajectories are (laevorotatory) ellipses.

13.8 Impairment Due to Waves

Can a wave $u(x,t) = \varphi(x-ct)$ induce any damage to a material? Assuming that the damage depends on the strain ε, we obtain with $\varepsilon = \partial u/\partial x = \varphi'$ and with $v = \partial u/\partial t = -c\varphi'$ the result $|\varepsilon| = v/c$. Since c is given for a considered medium, the criterion for damage is whether the velocity v (measured by so-called geophones) exceeds particular limits that are given in codes.

14 Plasticity Theory

14.1 Relevance of Plasticity Theory

For decades, plasticity theory (also called elastoplasticity theory) was considered as *the* theoretical tool to describe the inelastic deformation of solid materials. In the 1960s Roscoe and his colleagues at the University of Cambridge pioneered the application of plasticity theory in soil mechanics. Since then, an immense diversity of (elasto)plastic constitutive equations have been derived for soil. Their presentation in this book is intentionally kept short, as the author believes that the new theory of barodesy (Chapter 16) offers a better way to understand and describe the behaviour of soil.

14.2 One-dimensional Origin of Plasticity

In one-dimensional (1D) bodies (e.g. a string), plastic strain (yield) sets in when the stress σ exceeds a particular yield limit, i.e. if $\sigma > \sigma_Y$. In three dimensions, however, an inequality between tensors is meaningless. To still operate with inequalities, scalar functions of the stress tensor have to be introduced, so that we can write e.g. $f(\sigma_{ij}) < 0$. This gave rise to the introduction of the yield function $f(\sigma_{ij}, \dots)$ and inequalities as shown in the next section.

14.3 Yield Function, Loading–Unloading

According to plasticity theory, deformation is split into an elastic and a plastic part: $\varepsilon_{ij} = \varepsilon_{ij}^e + \varepsilon_{ij}^p$. The yield function $f(\sigma_{ij}, \varepsilon_{ij}^p)$ is introduced so that the equation $f = 0$ is the yield surface in stress space, which encloses the 'elastic range'. In the case of 'ideal plasticity', f does not depend on ε_{ij}^p, while the dependence of f on ε_{ij}^p constitutes 'hardening'. With the help of the yield function, loading can be defined as follows:

$$f = 0 \quad \text{and} \quad \frac{\partial f}{\partial \sigma_{ij}} \mathrm{d}\sigma_{ij} > 0,$$

whereas unloading prevails for

$$f < 0$$

$$\text{or} \qquad f = 0 \qquad \text{and} \qquad \frac{\partial f}{\partial \sigma_{ij}} \mathrm{d}\sigma_{ij} < 0.$$

The case $\frac{\partial f}{\partial \sigma_{ij}} \mathrm{d}\sigma_{ij} = 0$ is called neutral loading. At loading, ε^p_{ij} is variable, i.e. $\mathrm{d}\varepsilon^p_{ij} \neq 0$, and the 'consistency condition'

$$\mathrm{d}f = \frac{\partial f}{\partial \sigma_{ij}} \mathrm{d}\sigma_{ij} + \frac{\partial f}{\partial \varepsilon^p_{ij}} \mathrm{d}\varepsilon^p_{ij} = 0 \tag{14.1}$$

implies that at loading a moving stress point also moves the yield surface, i.e. it drags it behind so that it always lies upon it. The direction of the plastic strain increment $\mathrm{d}\varepsilon^p_{ij}$ is given by the 'plastic potential' $g(\sigma_{ij})$:

$$\mathrm{d}\varepsilon^p_{ij} = \lambda \frac{\partial g}{\partial \sigma_{ij}}. \tag{14.2}$$

Equation 14.2 is the 'flow rule'. λ can be obtained by introducing Equation 14.2 into Equation 14.1:

$$\lambda = -\frac{\dfrac{\partial f}{\partial \sigma_{kl}}}{\dfrac{\partial f}{\partial \varepsilon^p_{pq}} \dfrac{\partial g}{\partial \sigma_{pq}}} \mathrm{d}\sigma_{kl}.$$

Thus, we have for loading

$$\mathrm{d}\varepsilon_{ij} = \mathrm{d}\varepsilon^e_{ij} + \mathrm{d}\varepsilon^p_{ij} = \left[E_{ijkl} - \frac{\frac{\partial f}{\partial \sigma_{kl}} \frac{\partial g}{\partial \sigma_{ij}}}{\frac{\partial f}{\partial \varepsilon^p_{pq}} \frac{\partial g}{\partial \sigma_{pq}}} \right] \mathrm{d}\sigma_{kl} \tag{14.3}$$

and for unloading elastic behaviour, i.e.

$$\mathrm{d}\varepsilon_{ij} = E_{ijkl} \mathrm{d}\sigma_{kl}. \tag{14.4}$$

For $f \neq g$ the matrix in brackets [...] in Equation 14.3 is not symmetric. According to Equations 14.3 and 14.4 the elastoplastic response is composed of two linear relations, one that holds for loading and one for unloading.

The fact that shear strength increases with hydrostatic pressure (due to friction) implies that the yield surface inflates with increasing hydrostatic pressure and is therefore often depicted as a cone. The directrix of this cone does not necessarily have to be a circle. Since a cone is an open surface, the elastic range has to be closed by a so-called cap which expands with volume reduction (compression).

14.4 Normality Rule

The special case where $f = g$ is called 'normality rule' or 'associated flow rule'. It is the basis for several theorems, so it was long considered a fundamental material property. However, this is not the case, e.g. the normality rule does not apply for sand, because this would imply that the friction angle φ is equal to the dilatancy angle ψ, whereas measurements and theoretical considerations show that $\varphi > \psi$ applies.

The normality rule can be derived from Drucker's 'stability criterion': $\dot{\sigma}_{ij}\dot{\varepsilon}^{pl}_{ij} \geq 0$. Materials that fail to satisfy this criterion are associated with difficulties (i.e. non-uniqueness or singularities) in numerical solutions based on plasticity theory.

14.5 Collapse or Limit Load Theorems

Let us look at the estimates of the bearing capacity of a strip footing in frictionless soil (from Chapter 10). The three methods considered (see Equations 10.12, 10.13, 10.14) yield the following approximations for the limit load p_L:

$$
\begin{array}{ll}
\text{slip circle:} & p_L = 5.5\ c, \\
\text{stress discontinuity:} & p_L = 4\ c, \\
\text{slip lines (Prandtl):} & p_L = 5.14\ c.
\end{array}
$$

In view of this diversity, one wonders what the true limit load is. The so-called collapse theorems of plasticity theory make it possible to limit the true limit (or ultimate) load on the basis of the above-mentioned solutions. The terms 'statically admissible stress field' and 'kinematically admissible velocity field' are needed to introduce this approach. A statically admissible stress field fulfils static boundary conditions and equilibrium conditions, and it does not violate the condition $f(\sigma_{ij}) \leq 0$. The equation $f = 0$ characterises the limit state. For example, the limit condition for a frictional material according to Mohr–Coulomb is given by Equation 9.1 or $f = (\sigma_1 + \sigma_2)\sin\varphi - (\sigma_1 - \sigma_2) = 0$. A kinematically admissible velocity field satisfies the kinematic boundary conditions as well as any internal constraints (e.g. volume constancy). In connection with the collapse theorems, a collapse mechanism can be defined as a kinematically admissible velocity field.

The collapse theorems are:

Static collapse theorem: A body does not collapse if there exists at least one admissible stress field.

Reversal of the static collapse theorem: A body collapses if there is no admissible stress field for it.

Kinematic collapse theorem: A body fails when there is a collapse mechanism in which the power $\mathcal{A}$ of the external forces (e.g. weight and surface loads) exceeds the dissipation power $\mathcal{D}$ needed to overcome the shear strength, i.e. if $\mathcal{A} > \mathcal{D}$, where

$$
\begin{aligned}
\mathcal{A} &= \mathcal{A}(\boldsymbol{X},\boldsymbol{p}) = \int_V \boldsymbol{X}\cdot\boldsymbol{v}\,\mathrm{d}V + \int_S \boldsymbol{p}\cdot\boldsymbol{v}\,\mathrm{d}S, \\
\mathcal{D} &= \int_V \sigma^l_{ij}\dot{\varepsilon}_{ij}\,\mathrm{d}V + \int_{S_0} \sigma^l_{ij} n_j \delta v_i \,\mathrm{d}S_0
\end{aligned}
$$

with

$\boldsymbol{X}$: volume force (e.g. gravitation)

$\boldsymbol{v}$: velocity

$\boldsymbol{p}$: pressure on the surface of the body

V: volume

S: surface

S_0: surface of internal discontinuities
n_j: unit vector normal on discontinuities
δv_i: velocity jump on discontinuities
$\dot{\varepsilon}_{ij}$: $\frac{1}{2}\left(\frac{\partial v_i}{\partial x_j}+\frac{\partial v_j}{\partial x_i}\right)$
σ^l_{ij}: stress that satisfies the limit condition $f(\sigma_{ij}) = 0$, not necessarily the actual stress. To achieve this, the external loads $\boldsymbol{X}$ and $\boldsymbol{p}$ must be increased to the fictitious values $\boldsymbol{X}^*$ and $\boldsymbol{p}^*$, such that $\mathcal{A}(\boldsymbol{X}^*,\boldsymbol{p}^*) = \mathcal{D}$.

Reversal of the kinematic collapse theorem: A body does not fail if for all conceivable collapse mechanisms, $\mathcal{A} < \mathcal{D}$ holds.

Remark 1. The collapse theorems allow statements on the collapse of a body without knowledge of the real stress or velocity fields. They only consider 'admissible' fields, which do not violate the boundary, equilibrium and limit conditions, but are otherwise freely selectable and therefore relatively easy to determine. For the application of the collapse theorems (not their inversions), it is sufficient in each case to find one statically or kinematically admissible field. The inversions of the collapse theorems require the investigation of *all* (i.e. infinitely many) admissible fields, they are therefore by far not as helpful as the collapse theorems themselves. For practical applications, however, one does not need to examine an infinite number of fields. With a little experience, the examination of only a few fields leads to a useful result.
2. The validity of the collapse theorems is linked to a material property that characterises the plastic flow (i.e. the deformation in the limit state) and is called the 'normality condition', see Section 14.4.
3. The static collapse theorem is also called 'lower bound theorem', because it allows to limit the collapse load from below (i.e. from the safe side). According to this collapse theorem, a material does 'its best' to carry the loads imposed on it.
4. The kinematic collapse theorem is also called the *upper bound theorem*, because it allows the limit load to be bounded from above (i.e. from the unsafe side). It states that collapse will occur if it can occur. The application of the kinematic collapse theorem is particularly simple when considering collapse mechanisms composed of rigid blocks (Section 10.4).

Consider the bearing capacity of a strip footing, the collapse (or limit) load of $p_L = 5.5c$ is obtained using the kinematic collapse theorem, and represents thus an upper bound for the real collapse load. The solutions $p_L = 5.14c$ and $p_L = 4c$ were determined from stress fields and represent thus lower bounds for the collapse load.

14.6 Elastoplastic Relations for Soil

Some common elastoplastic constitutive equations for soil are:

Cam-clay: The Cam-clay theory is the first elastoplastic model designed for soil (normal-to-light over-consolidated clay). Initially it was only used to interpret the results of triaxial experiments, therefore in the original versions only the deformation variables $\varepsilon_q := \frac{2}{3}(\varepsilon_1 - \varepsilon_3)$ and $\varepsilon_v := \varepsilon_1 + 2\varepsilon_3$ appear along with the stress variables $q := \sigma_1 - \sigma_3$ and $p := \frac{1}{3}(\sigma_1 + 2\sigma_3)$. The factors $\frac{2}{3}$ and $\frac{1}{3}$ are to make the expression $p\varepsilon_v + q\varepsilon_q$ identical with $\sigma_1\varepsilon_1 + \sigma_2\varepsilon_2 + \sigma_3\varepsilon_3$ (= deformation work per unit volume). Here, ε_1 is the strain in the axial direction and $\varepsilon_2 \equiv \varepsilon_3$ is the strain in the radial direction. The transition to general stress and deformation tensors is achieved by interpreting the above-mentioned variables as invariants:

$$\varepsilon_v := \varepsilon_{kk} = \varepsilon_{11} + \varepsilon_{22} + \varepsilon_{33}, \quad p = \frac{1}{3}\sigma_{kk} = \frac{1}{3}(\sigma_{11} + \sigma_{22} + \sigma_{33}),$$

$$\varepsilon_q := \frac{2}{\sqrt{6}}\sqrt{\varepsilon^*_{ik}\varepsilon^*_{ki}} \quad \text{with} \quad \varepsilon^*_{ij} = \varepsilon_{ij} - \frac{1}{3}\varepsilon_{kk}\delta_{ij},$$

$$q := \sqrt{\frac{3}{2}\sigma^*_{ik}\sigma^*_{ki}} \quad \text{with} \quad \sigma^*_{ij} = \sigma_{ij} - \frac{1}{3}\sigma_{kk}\delta_{ij}.$$

The associated flow rule is used, i.e. equality of the yield surface and plastic potential is assumed, $f = g$. The cone portion of the yield surface is given by $q = Mp$, where M is related to the friction angle φ_c at the critical state with $M = 6\sin\varphi_c/(3 - \sin\varphi_c)$. The cap is given by an ellipse in the q-p space:

$$q^2 - M^2[p(p_c - p)] = 0,$$

where p_c is the consolidation stress (for hydrostatic consolidation) and is referred to as the strain-hardening parameter by relating it to the plastic volumetric strain by

$$\dot{p}_c = p_c \cdot \frac{1+e}{\lambda - \kappa} \cdot \dot{\varepsilon}^p_v,$$

where λ and κ are material constants that can be adjusted by adapting the initial hydrostatic loading and unloading to the relations $e = e_0 - \lambda \ln(p/p_0)$ and $e = e_1 - \kappa \ln(p/p_1)$, respectively. Within the yield surface only 'elastic' deformations take place:

$$\dot{\varepsilon}^e_v = \frac{\kappa}{(1+e)p}\dot{p}; \quad \dot{\varepsilon}^e_q = \frac{2}{9} \cdot \frac{1+\nu}{1-\nu} \cdot \frac{\kappa}{(1+e)p}\dot{q},$$

where ν is the Poisson number.

Elastoplasticity with the Mohr–Coulomb limit condition:

The cone part of the yield surface is given by

$$(\sigma_{max} - \sigma_{min}) = (\sigma_{max} + \sigma_{min}) \cdot \sin\varphi + 2c\cos\varphi,$$

where σ_{max} and σ_{min} are the maximum and minimum principal stresses. A cap is not defined, the constitutive equation is therefore incomplete. Consequently, for oedometric loading and unloading elastic behaviour is predicted.

14.7 Criticism of Plasticity Theory in Soil Mechanics

To treat tensors, such as stress and strain, with the aid of scalar functions of their invariants has been considered the only way to introduce constitutive equations for irreversible deformation. The cornerstone of plasticity theory is the yield surface in the stress space that encompasses the elastic range. Apart from the fact that the resulting stress–strain relation is hidden in the geometry of the yield surface and altogether complicated and difficult to handle, both conceptually and numerically, it should be taken into account that soils have no elastic range.

14.7.1 Remarks by Palmer and Pearce

In 1973 Palmer and Pearce published a paper titled 'Plasticity theory without yield surfaces' [77]. Some extracts from this paper support the criticism of the 'classical' plasticity theory and forecast, in some way, later approaches such as hypoplasticity and barodesy:

> It was quite natural that the idea of a yield surface should assume such importance in a theory built on experience with metals, since in most metals yield occurs at a fairly well-defined stress level. ...
>
> In soil mechanics the status of the yield surface concept is quite different, both in theory and experiment...
>
> ...strain measurements in clay depend on direct observation of boundary displacements, so that only quite large strain increments are reliably measurable, creep and pore-pressure diffusion confuse results...
>
> ...yield surface motions during strain-hardening are often too complex for the results to be helpful in constructing usable stress-strain relations.
>
> Might it be possible to resolve this (dilemma) by constructing a different kind of plasticity model, in which the yield surface concept had been dropped or relegated to a minor role?
>
> ...it might be useful to idealise clay as a material in which the yield surface has shrunk to a point, so that all deformations are plastic and *any* changes of stress from the current state will produce plastic strain increments.
>
> Palmer and Pearce [77]

The work of Palmer and Pearce referred to clay, but their arguments are equally valid for sand. Based on two postulates by Ilyushin, they proposed a plasticity theory without yield surfaces:

> The deviatoric stress has two components. The magnitude of the first component is a function of the octahedral shear strain, and its direction coincides with the principal strain vector (referring strain to an isotropically-consolidated initial state). The magnitude of the second component is constant, and its direction coincides with the current strain rate ...
>
> Reversal of the strain path would reverse the second component but not the first ...
>
> Palmer and Pearce [77]

The very last sentence strongly resembles a basic concept of hypoplasticity and barodesy, to which presumably the authors would have concluded had they used rate equations instead of finite ones.

15 Hypoplasticity

15.1 Hypoplasticity as an Alternative to Elastoplasticity

Elastoplasticity uses the so-called yield function $f(\sigma_{ij})$ such that stress increments $\Delta\sigma_{ij}$ leading outside the surface $f(\sigma_{ij}) = 0$ are considered as 'loading', otherwise as 'unloading'. This surface embraces the 'elastic range'. However, soil has no elastic range, and hence, this concept becomes questionable. So, the question arises as to how to describe irreversible deformation without using a yield function? A basic step towards this goal is to define different stiffnesses for loading and unloading, say $\Delta\sigma_{ij}=C^1_{ijkl}\Delta\varepsilon_{kl}$ for loading and $\Delta\sigma_{ij}=C^2_{ijkl}\Delta\varepsilon_{kl}$ for unloading. But, how to distinguish loading from unloading? Hypoplasticity offers an answer without resorting to yield functions.

15.2 Non-linear Rate Equations

A constitutive equation $\dot\sigma_{ij} = \dot\sigma_{ij}(\sigma_{kl}, \dot\varepsilon_{mn})$, which is linear in $\dot\varepsilon_{mn}$, can also be represented in the form $\dot\sigma_{ij} = M_{ijmn}\dot\varepsilon_{mn}$, where the stiffness matrix M_{ijmn} can depend on σ_{kl}. In the case of irreversible deformation, the stiffness also depends on the direction of the deformation, i.e. on $\dot\varepsilon^0_{mn} := \dot\varepsilon_{mn}/|\dot\varepsilon_{mn}|$. The simplest extension of a constitutive equation towards irreversible deformation is therefore

$$\dot\sigma_{ij} = (M_{ijmn} + N_{ij}\dot\varepsilon^0_{mn})\dot\varepsilon_{mn}. \tag{15.1}$$

Here, M_{ijmn} and N_{ij} may depend on the stress σ_{ij}. Equation 15.1 can also be written in the following form:

$$\dot\sigma_{ij} = M_{ijmn}\dot\varepsilon_{mn} + N_{ij}|\dot\varepsilon_{mn}|,$$

where $|\dot\varepsilon_{mn}| = \sqrt{\dot\varepsilon_{mn}\dot\varepsilon_{mn}}$ is the norm of $\dot\varepsilon_{mn}$. The term $M_{ijmn}\dot\varepsilon_{mn}$ represents a tensor-valued function of σ_{kl} and $\dot\varepsilon_{mn}$ which is linear in $\dot\varepsilon_{mn}$. The term N_{ij} represents a tensor-valued function of σ_{kl}.

There are various versions of hypoplastic constitutive equations that emanated from consecutive improvements of existing versions [53]. Without the burden of yield functions, plastic potentials and their evolutions in the stress space, hypoplastic constitutive equations are characterised by simplicity, which is also reflected in their calibration and finite element (FEM) implementation. In particular, hypoplastic constitutive equations do not refer to any elastic range, which does not exist for soils.

15.3 Notation

In the literature on hypoplasticity, the so-called symbolic notation is often used instead of component notation: $\mathbf{T}=\sigma_{ij}$, $\mathbf{D}=\dot{\varepsilon}_{ij}$, $\sqrt{\mathrm{tr}\mathbf{D}^2}=|\dot{\varepsilon}_{mn}|$. Actually, the stress rate is not $\dot{\mathbf{T}}$, but $\overset{\circ}{\mathbf{T}}$, because $\dot{\mathbf{T}}$ is not an objective quantity, whereas $\overset{\circ}{\mathbf{T}}$ represents an objective stress rate (see Section 18.2). The difference between $\dot{\mathbf{T}}$ and $\overset{\circ}{\mathbf{T}}$ is numerically small, so it will not be pursued here, although it is the subject of much discussion.

The general mathematical equation for hypoplasticity has the form

$$\overset{\circ}{\mathbf{T}} = \mathbf{L}(\mathbf{T})\mathbf{D} + \mathbf{N}(\mathbf{T})|\mathbf{D}| = (\mathbf{L} + \mathbf{N} \otimes \mathbf{D}^0)\mathbf{D}, \tag{15.2}$$

where $\mathbf{L}$ is a linear operator that depends on $\mathbf{T}$ and is applied to $\mathbf{D}$, and $\mathbf{N}$ is a tensor-valued function of $\mathbf{T}$. If $\mathbf{L}$ is considered as a fourth-order tensor, then the notation $\overset{\circ}{\mathbf{T}}= \mathcal{L} : \mathbf{D} + \mathbf{N}|\mathbf{D}|$ can be used.

15.4 Incremental Non-linearity

The step from hypoelasticity to hypoplasticity consists in abandoning the requirement of $\mathbf{h}(\mathbf{T},\mathbf{D})$ to be linear in $\mathbf{D}$ and replacing it by the weaker requirement of $\mathbf{h}(\mathbf{T},\mathbf{D})$ to be homogeneous in the first degree but *non-linear* in $\mathbf{D}$. This makes it possible to describe irreversible material behaviour and constitutes the so-called *incremental non-linearity*, as $\mathbf{h}(\mathbf{T},-\mathbf{D}) \neq -\mathbf{h}(\mathbf{T},\mathbf{D})$.

Incremental non-linearity contradicts the misconception that, in physics, *every* relation is linear *in the small*. This means that a relation $y(x)$ can be linearised at $x = x_0$ within a small neighbourhood of x_0: $y \approx a(x - x_0)$ with $a = y'(x_0)$. However, this is not the case with materials undergoing irreversible deformation.

15.5 Mathematical Description of Irreversibility

Constitutive equations are relations between stress σ_{ij} and strain ε_{kl}. The irreversible character of soil deformation does not allow us to define these relations as functions $\sigma_{ij} = \sigma_{ij}(\varepsilon_{kl})$ or $\varepsilon_{kl} = \varepsilon_{kl}(\sigma_{ij})$. This would imply that σ_{ij} (or ε_{kl}) does *not* depend on the history of ε_{kl} (or σ_{ij}).

The influence of history is an essential characteristic of the deformation behaviour of soil, which is characterised, for example, by footprints in the sand (the deformation remains, although the load has passed).

One way to mathematically describe history-dependent relationships is to represent σ_{ij} not as function of ε_{kl} but to relate the corresponding increments $\mathrm{d}\sigma_{ij}$ with $\mathrm{d}\varepsilon_{kl}$, i.e. to use so-called differential (or Pfaffean) forms. A relationship $\mathrm{d}\sigma_{ij} = \mathrm{d}\sigma_{ij}(\mathrm{d}\varepsilon_{kl})$ is then history-dependent, if it is *not integrable*.

Example

Consider the following relation between x and y:

$$dy = dx + |dx| = \begin{cases} 2dx & \text{for } dx > 0 \\ 0 & \text{for } dx < 0 \end{cases}, \tag{15.3}$$

and the following histories of x:

- *History 1:* x increases from $x = 0$ to $x = 5$ ($dx > 0$).
- *History 2:* x increases from $x = 0$ to $x = 10$ ($dx > 0$) and then reduces from $x = 10$ to $x = 5$ ($dx < 0$).

Starting from $y(x = 0) = 0$ we get the value $y(x{=}5){=}10$ for History 1 and $y(x = 5) = 20$ for History 2.

As stated, history-dependence requires that the constitutive equations for soil are not finite, i.e. $\sigma_{ij} = \sigma_{ij}(\varepsilon_{kl})$, but incremental, i.e. $d\sigma_{ij} = d\sigma_{ij}(d\varepsilon_{kl})$. As an alternative to the increments (differentials) $d\sigma_{ij}$ and $d\varepsilon_{kl}$ one can use the rates $\dot{\sigma}_{ij} = d\sigma_{ij}/dt$ and $\dot{\varepsilon}_{kl} = d\varepsilon_{kl}/dt$. Constitutive equations of the type $\dot{\sigma}_{ij} = \dot{\sigma}_{ij}(\dot{\varepsilon}_{kl})$, are called rate or evolution equations.

The dependence of stiffness $\partial\sigma_{ij}/\partial\varepsilon_{kl}$ (or $\dot{\sigma}_{ij}/\dot{\varepsilon}_{kl}$) on stress σ_{ij} leads to constitutive equations of the type $\dot{\sigma}_{ij} = \dot{\sigma}_{ij}(\sigma_{kl}, \dot{\varepsilon}_{mn})$. The irreversibility of soil deformation implies that the stiffness $\dot{\sigma}_{ij}/\dot{\varepsilon}_{kl}$ does not have the same value at loading and unloading, i.e.

$$\dot{\sigma}_{ij}(\sigma_{kl}, \dot{\varepsilon}_{mn}) \neq -\dot{\sigma}_{ij}(\sigma_{kl}, -\dot{\varepsilon}_{mn}),$$

and this means that the function $\dot{\sigma}_{ij}(\sigma_{kl}, \dot{\varepsilon}_{mn})$ is non-linear in $\dot{\varepsilon}_{mn}$ (so-called incremental non-linearity). Elastoplasticity and hypoplasticity comply with this rule.

15.6 Emergence of Hypoplasticity

We consider a rate equation $\overset{\circ}{\mathbf{T}} = \mathbf{h}(\mathbf{T}, \mathbf{D})$. According to a representation theorem, a function $\mathbf{h}(\mathbf{T}, \mathbf{D})$ can be represented as

$$\begin{aligned}\mathbf{h}(\mathbf{T}, \mathbf{D}) &= a_1\mathbf{1} + a_2\mathbf{T} + a_3\mathbf{D} + a_5\mathbf{T}^2 + a_6\mathbf{D}^2 \\ &\quad + a_7\,\mathrm{sym}(\mathbf{TD}) + a_8\,\mathrm{sym}(\mathbf{T}^2\mathbf{D}) + a_9\,\mathrm{sym}(\mathbf{TD}^2) + a_{10}\,\mathrm{sym}(\mathbf{T}^2\mathbf{D}^2),\end{aligned}$$

where $\mathrm{sym}(\mathbf{TD})$ means $(\mathbf{TD} + \mathbf{DT})/2$. A paradigm for such a function is Truesdell's theory of hypoelasticity where, however, the function $\mathbf{h}(\mathbf{T}, \mathbf{D})$ is *linear* in $\mathbf{D}$. Hence, the same stiffness for loading and unloading is predicted.

Hypoplasticity uses a function $\mathbf{h}(\mathbf{T}, \mathbf{D})$, which is homogeneous in the first degree in $\mathbf{D}$ (to assure rate independence) but *non-linear* in $\mathbf{D}$. This is, for example, the case with the term $\dfrac{\mathbf{D}^2}{\sqrt{\mathrm{tr}\mathbf{D}^2}}$. The remainder of the formulation of a hypoplastic constitutive equation was trial and error and led in 1977 to the first version by the author, which did not yet bear the name hypoplasticity. This early version was clumsy but was, in a sense, a success, as it proved that some features of soil can – in principle – be described without resorting to the theory of plasticity. The main shortcomings of the first version were:

- it was homogeneous of the zeroth degree in $\mathbf{T}$, which implies the unrealistic result that an increase in stress level will leave the stiffness unaltered.
- the density or void ratio was not included. It was expected that the whatsoever obtained material constants are valid for, say, dense sand, whereas for loose sand other constants had to be used.

The following important insights were gained:

- an alternative to plasticity theory was launched
- the mathematical structure of the equation and the underlying physics became more evident:
  - the homogeneity of the first degree in $\mathbf{D}$ is linked to rate independence
  - the degree of homogeneity in $\mathbf{T}$ describes how stiffness depends on stress level.

A step forward was achieved in 1984 by the author proposing a much simpler equation [43]:

$$\overset{\circ}{\mathbf{T}} = \frac{c_1}{2}(\mathbf{TD} + \mathbf{DT}) + c_2 \mathrm{tr}\,(\mathbf{TD}) \cdot \mathbf{1} + \left(c_3 \mathbf{T} + \frac{c_4 \mathbf{T}^2}{\mathrm{tr}\,\mathbf{T}} \right) |\mathbf{D}|. \tag{15.4}$$

This equation is homogeneous in the first degree with respect to $\mathbf{T}$, which implies that the stiffness is proportional to $|\mathbf{T}|$. The further development of hypoplasticity is outlined in [53].

15.7 Links to Elastoplasticity

In 1991 the author pointed to the possibility of expressing the limit surface in stress space by means of the hypoplastic equation [44]. Transforming the general hypoplastic equation $\overset{\circ}{\mathbf{T}} = \mathbf{L}(\mathbf{T}, \mathbf{D}) + \mathbf{N}(\mathbf{T})|\mathbf{D}|$ to $\overset{\circ}{\mathbf{T}} = \mathcal{L}(\mathbf{T})[\mathbf{D} + \mathbf{B}(\mathbf{T})|\mathbf{D}|]$ with $\mathbf{B}(\mathbf{T}) = \mathcal{L}^{-1}\mathbf{N}$ and considering the limit state $\overset{\circ}{\mathbf{T}} = \mathbf{0}$ leads to $\mathbf{D}^0 + \mathbf{B} = \mathbf{0}$, with $\mathbf{D}^0 := \mathbf{D}/|\mathbf{D}|$. Hence, the limit state can be expressed by the equation $|\mathbf{B}| = 1$. As $\mathbf{B}$ depends only on stress, this equation can be seen as a yield surface $f(\sigma_{ij}) = 0$ in the terminology of plasticity theory. For example, the Drucker–Prager yield surface is expressed by $\left|\frac{\mathbf{T}^*}{\mathrm{tr}\mathbf{T}}\right| = \mathrm{const}$ The resulting equation $\mathbf{D}^0 := \mathbf{D}/|\mathbf{D}| = -\mathbf{B}(\mathbf{T})$ can be seen as the corresponding flow rule.

15.8 Improving Memory by Means of Intergranular Strain

The intergranular strain introduced in 1997 by Niemunis and Herle [75] made it possible to remedy a principal shortcoming of hypoplasticity, the excessive accumulation of deformation at small stress cycles (so-called ratcheting), and thus, to model cyclic loading and the so-called small strain effects.

The underlying idea was that grains are connected with bonds, which are first deformed, and upon continuation of the deformation, they yield such that further deformation occurs with grain rearrangement. The deformation of these bonds is called 'intergranular strain' $\boldsymbol{\delta}$. The intergranular strain tensor $\boldsymbol{\delta}$ should not be confused with the dilatancy δ (Section 16.2).

16 Barodesy

16.1 Introduction

Barodesy, launched by the author in 2012, is a constitutive theory for soil and, in general, for granular solids. As the name signifies, it has nothing to do with either elasticity or with plasticity. It is characterised by a straightforward derivation from simple principles stating, virtually, that dry sand cannot sustain tensile stresses and has a fading memory.

The version presented here differs from previous ones [47, 49–53] by a new approach to critical void ratio, firstly presented in [54].

16.2 Notation

Some obvious symbols are used here that help to make barodesy more concise and easier to grasp. As usual in mechanics, σ denotes the stress and ε denotes the strain. Furthermore,

the stress variables p **and** q are used for axisymmetric (triaxial) stress states $p = (\sigma_1 + 2\sigma_2)/3$, $q = \sigma_1 - \sigma_2$.

$e = V_v/V_s$: void ratio, e_c: critical void ratio. V_v and V_s are the volumes of voids and solids (grains), respectively, in a small but representative volume element.

value of a tensor $\mathbf{A}$ is its Frobenius norm: $|\mathbf{A}| := \sqrt{\mathrm{tr}\mathbf{A}^2}$.

normalisation of a tensor $\mathbf{A}$ is denoted by the exponent 0, i.e. $\mathbf{A}^0 := \mathbf{A}/|\mathbf{A}|$. $\mathbf{A}^0$ is a unit tensor, i.e. $|\mathbf{A}^0| = 1$. A tensor can be represented by its value and the corresponding unit tensor, e.g. $\mathbf{T} = \sigma\mathbf{T}^0, \mathbf{D} = \dot{\varepsilon}\mathbf{D}^0$.

$\sigma := |\mathbf{T}|$, the value of stress $\mathbf{T}$.

$\dot{\varepsilon} := |\mathbf{D}|$, the value of stretching $\mathbf{D}$.

$\delta := \mathrm{tr}\mathbf{D}^0 = \mathrm{tr}\mathbf{D}/|\mathbf{D}|$, ratio of volume change to total deformation rates. δ is a measure of dilatancy.

$\dot{\mathbf{T}}$: time rate of stress. For simplicity, objective stress rates $\overset{\circ}{\mathbf{T}}$ are not addressed here, assuming $\overset{\circ}{\mathbf{T}} \approx \dot{\mathbf{T}}$. Strictly speaking, an objective constitutive relation yields $\overset{\circ}{\mathbf{T}}$ rather than $\dot{\mathbf{T}}$.

$\dot{\lambda} := |\frac{\mathrm{d}}{\mathrm{d}t}\mathbf{D}^0|$, this scalar indicates the value of the time derivative of $\mathbf{D}^0$. For $\mathbf{D}^0 =$ const we have $\dot{\lambda} = 0$, see Section 16.5.3.

$c_1, \ldots, c_5, k_1, k_2, \kappa_1, \kappa_2$: material constants. It facilitates reading to clearly denote the material constants as such, so as to distinguish them from quantities that depend on other variables.

proportional paths of stress and strain are straight lines through the origins of the corresponding three-dimensional (3D) spaces of principal stresses and principal strains. They are characterised by $\sigma_1 : \sigma_2 : \sigma_3 = \text{const}$ and $\varepsilon_1 : \varepsilon_2 : \varepsilon_3 = \text{const}$, respectively.

16.3 Derivation of the Constitutive Equation

The constitutive equation of barodesy can be more or less derived from two rules formulated in 1976 by Goldscheider [30, 31], see Section 2.8.1, which refer to proportional paths.

16.3.1 Implications of Proportional Paths

We consider first the simplest possible deformation histories, which are proportional rectilinear extensions: $\mathbf{E} = \alpha \mathbf{D}^0$, with α being a monotonically increasing function of time, $\alpha(0) = 0$. In the geometrical sense, proportional strain paths are straight lines in the 3D space of principal strains, emanating from the origin $\mathbf{E} = \mathbf{0}$. The unit tensor $\mathbf{D}^0$ expresses the direction of a proportional strain path. For these simple strain histories the stress response $\mathbf{T}$ is a function of strain: $\mathbf{T} = \mathbf{f}(\mathbf{E})$, and it appears reasonable to assume that the stress response to proportional straining is a proportional stress path too, $\mathbf{T} = \beta \mathbf{R}^0$, where β is an increasing function of time, $\beta(0) = 0$.

16.3.2 Proportional Paths for Cohesionless Granulates

The statements made so far are quite general and will possibly apply to any solid material. The specialties of soil start with the fact that proportional stress paths must be restricted to compressions, as no tensile stress can be transmitted by the grains. This fact affects the mapping of proportional strain paths to the corresponding proportional stress paths, i.e. the function $\mathbf{R}^0(\mathbf{D}^0)$, which maps the direction $\mathbf{D}^0$ of a proportional strain path to the direction $\mathbf{R}^0$ of the corresponding stress path. As the grains (here assumed as rigid) of a cohesionless soil cannot transmit tensile forces, only volume-reducing or volume-preserving straining is feasible, and the corresponding stresses must not have tensile principal values. Consequently, the function $\mathbf{R}(\mathbf{D}^0)$ maps volume-reducing (i.e. $\mathrm{tr}\mathbf{D}^0 \leq 0$) proportional strain paths onto proportional stress paths which are within the compressive octant of the principal stress space. This is achieved by the equation $\mathbf{R}(\mathbf{D}) = -\exp\left(c_1 \mathbf{D}^{0\star}\right) + c_2 \delta \mathbf{1}$, a discussion of which in given in Section 16.6. As for volume-*increasing* proportional strain paths, they are still possible provided that they start at a non-vanishing stress state. Even without experiments, we can reason that the corresponding stress paths head finally to the stress-free state $\mathbf{T} = \mathbf{0}$.

With Equation 16.7 we can write the constitutive relation $\mathbf{T} = \beta(t)\, \mathbf{R}^0$ which is only valid for proportional strain paths emanating from $\mathbf{T} = \mathbf{0}$. If proportional straining starts at a deformation-induced stress $\mathbf{T}_0 \neq \mathbf{0}$, then a possible constitutive equation could be $\mathbf{T} = \beta(t)\, \mathbf{R}^0 + \mathbf{T}_0$. However, the principle of fading memory requires that the influence of $\mathbf{T}_0$ fades with increasing deformation. Hence, it appears reasonable to let the influence of $\mathbf{T}_0$ decrease with deformation (or with time, if deformation increases with time). This can be achieved by writing

$$\mathbf{T} = \beta(t)\, \mathbf{R}^0 + \gamma(t)\, \mathbf{T}_0, \tag{16.1}$$

with $\gamma(t)$ being a decreasing function of time, $\gamma(0) = 1$ and $\gamma(t \to \infty) = 0$. Differentiating (16.1) with respect to time and considering that the initial stress $\mathbf{T}_0$ equals at each time t the actual stress $\mathbf{T}$, yields the evolution equation

$$\dot{\mathbf{T}} = \dot{\beta}\, \mathbf{R}^0 + \dot{\gamma}\, \mathbf{T}, \tag{16.2}$$

which is the basic equation of barodesy. The final step is to realise that the assumptions met so far imply that this equation holds not only for proportional strainings and not only for rectlinear extensions but also for general deformations.

16.3.3 Fine Tuning

Fine tuning of the scalar quantities $\dot{\beta}$ and $\dot{\gamma}$ leads finally to Equation 16.6 (see Section 16.6.4). To this end, we set $\dot{\beta} = \hat{\beta}\dot{\varepsilon}$ and $\dot{\gamma} = (\hat{\gamma}/\sigma)\dot{\varepsilon}$ and obtain:

$$\dot{\mathbf{T}} = (\hat{\beta}\, \mathbf{R}^0 + \hat{\gamma}\, \mathbf{T}^0)\, \dot{\varepsilon}. \tag{16.3}$$

Now we consider limit states (either peak or residual), for which holds: $\dot{\mathbf{T}} = \mathbf{0}$. This tensorial equation implies $\mathbf{R}^0 = \mathbf{T}^0$ and $\hat{\beta} + \hat{\gamma} = 0$. Now we only need to assume that $\hat{\beta}$ and $\hat{\gamma}$ depend on dilatancy δ, on void ratio e and on critical void ratio e_c in such a way, that the sum $\hat{\beta} + \hat{\gamma}$ vanishes at peak and at residual states. This is achieved, for example, by setting:

$$\hat{\beta} = e_c + c_3\delta, \quad \hat{\gamma} = -e + c_3\delta, \tag{16.4}$$

because at residual (i.e. critical) states we have $\delta = 0$ and $e = e_c$, and at peak the equation $e_c - e + 2c_3\delta = 0$ is fulfilled with an appropriate value of c_3. To model the fact that the stiffness increases with σ it is set

$$\dot{\mathbf{T}} = h\, (\hat{\beta}\, \mathbf{R}^0 + \hat{\gamma}\, \mathbf{T}^0)\, \dot{\varepsilon} \quad \text{with} \quad h = c_4\sigma^{c_5}. \tag{16.5}$$

Herewith, the equations of barodesy are ready and summarised in the following section.

16.4 The Equations of Barodesy

Writing f for $\hat{\beta}$ and g for $\hat{\gamma}$, the 2020 version of barodesy reads:

$$\dot{\mathbf{T}} = h \cdot [f\mathbf{R}^0 + g\mathbf{T}^0] \cdot \dot{\varepsilon} \tag{16.6}$$

with

$$\mathbf{R}(\mathbf{D}) = -\exp\left(c_1 \mathbf{D}^{0\star}\right) + c_2 \delta \mathbf{1} \tag{16.7}$$

and

$$h = c_4 \sigma^{c_5}, \quad f = e_c + c_3 \delta, \quad g = -e + c_3 \delta. \tag{16.8}$$

16.4.1 Mathematical Structure

As in hypoplasticity and the previous versions of barodesy, the basic equation is an evolution equation of the type $\dot{\mathbf{T}} = \mathbf{h}(\mathbf{T}, \mathbf{D}, e, e_c)$. Compared with previous versions, the function $\mathbf{h}(.)$ now contains the critical void ratio e_c as an additional independent variable (or state parameter). As in previous versions, $\mathbf{h}(.)$ is non-linear in $\mathbf{D}$, and this makes it possible to describe irreversible (anelastic) behaviour. Still, $\mathbf{h}(\mathbf{T}, \mathbf{D}, e, e_c)$ is homogeneous in the first degree in $\mathbf{D}$, i.e. $\mathbf{h}(\mathbf{T}, \lambda\mathbf{D}, e, e_c) = \lambda\mathbf{h}(\mathbf{T}, \mathbf{D}, e, e_c)$ for $\lambda > 0$, and this implies that this constitutive equation describes a rate-independent material.

With $\Delta\sigma = \dot{\sigma}\Delta t$ and $\Delta\varepsilon = \dot{\varepsilon}\Delta t$, evolution equations yield the *change* of stress $\Delta\sigma$ due to a deformation $\Delta\varepsilon$. As a consequence, the knowledge of the initial stress is necessary. Its influence fades with increasing deformation.

16.5 Critical States Revisited

A basic innovation of this version of barodesy is the new concept of critical void ratio e_c. Related to the notions of dilatancy and contractancy, the concept of critical void ratio evolved gradually in soil mechanics. It was observed that the void ratio increases or decreases in the course of drained triaxial tests, depending on whether the considered sample was initially dense or loose, respectively. Of course, this increase or decrease must be limited, and the limit is called critical void ratio e_c.

In other words, the void ratio becomes stationary ($e \to e_c$) with continued drained (triaxial) deformation, as soon as a so-called critical state is reached. At this state also the stress becomes stationary, $\mathbf{T} = \text{const}$ upon further isochoric (i.e. volume preserving, $\text{tr}\mathbf{D} = 0$) deformation.

Critical stress states of triaxial tests fulfil the equation $q = mp$ (critical state line, CSL), with $m = \text{const}$ For non-axisymmetric conditions, the locus of critical stress states is the so-called critical state surface. The line $q = mp$ represents a proportional stress path pertaining to an isochoric ($\text{tr}\mathbf{D} = 0$) proportional strain path and serves as an attractor of stress paths of drained and undrained triaxial tests (Figs. 3.2 and 3.3). Constant volume characterises undrained triaxial tests and also the final (steady) stage of drained ones.

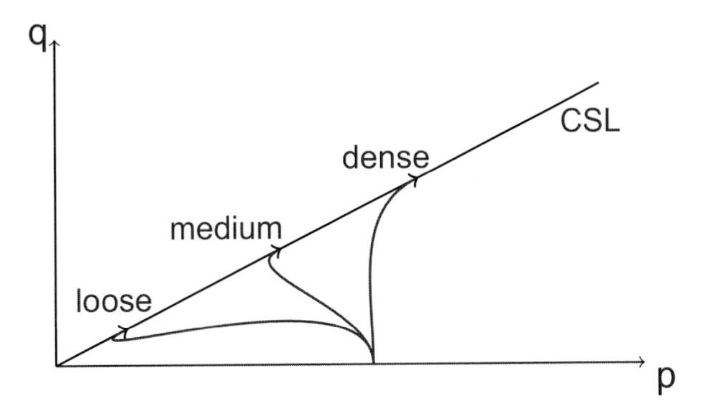

Figure 16.1 Stress paths corresponding to undrained triaxial tests (schematically, compare with experimental results in Fig. 2.12). They all approach the CSL.

16.5.1 Critical Void Ratio in Triaxial Tests

Triaxial tests are here understood as element tests. Typically, they start with an initial oedometric (or hydrostatic) consolidation, i.e. a proportional stress path. This path is followed

in drained triaxial tests, by a straight stress path (characterised either by $\sigma_2 = \sigma_3 =$ const or by $p =$ const). Depending on the initial void ratio, the stress point may overshoot the CSL and reach a so-called peak. Then it turns back ('softening') along the same stress path and finally ends up at the CSL (Fig. 3.3).

in undrained triaxial tests, by an isochoric deformation corresponding to a curved stress path (see the three examples in Fig. 16.1) that huddles asymptotically against the CSL and finally heads towards increasing p.

Referring to critical states, the common perception of test outcomes is, to some degree, biased by concepts of plasticity theory. Consider, in particular, the 'end points' of stress paths pertaining to isochoric deformation. According to plasticity theory, these stress paths end at the yield surface, whereas according to barodesy, they approach and then move along the CSL, see Figs. 16.1 and 16.4.

Stress paths of undrained dense clay often end with a peculiar hook-like sharp bend (Figs. 16.21 and 16.22) due to softening, which is related to loss of homogeneity of the deformation [3, 13, 37].

16.5.2 Critical States Along the Critical State Line

A stress state, represented as a point in stress space, can move along the CSL due to isochoric deformation up to a steady state, i.e. a state of constant deformation rate, constant stress and (necessarily) constant volume. This is the case when the stress point obtains that value of mean stress p, for which the actual void ratio e is critical. A relation $e_c = \epsilon_1(p)$ describes the dependence of e_c on p, that holds along the CSL $q = mp$. Note that the name CSL can be used for two different lines, (i) the stress path $q = mp$ and (ii) the relation $e_c = \epsilon_1(p)$, both of which are needed to describe critical states. $\epsilon_1(p)$ is a decreasing function of p and can be described, e.g. by

$$\epsilon_1(p) = k_1 - k_2 \ln(p), \tag{16.9}$$

where k_1 depends also on the dimension of p. Clearly, this simple relation is only valid within a limited range of p, as the function $\ln(p)$ is not bounded.

For the experimental determination of this relation, we need a separate triaxial test for each point of the CSL. Drained tests are carried out until the condition e = const is reached [8, 36], and undrained ones target the condition p = const (Figs. 28 in [40], 2 and 3 in [113]). In both cases the results are obscured by the onset of inhomogeneous deformation, which may also disturb the condition of constant volume: the generation of (hardly detectable!) thin dilating shear bands in clay enables a sort of 'internal drainage'.

16.5.3 Critical Void Ratio of Non-critical States

Following its definition, e_c can only be encountered at critical stress states ($q = mp$). However, e_c is important also at stress states far off the critical ones, as it serves to distinguish between dense ($e < e_c$) and loose ($e > e_c$) states. It therefore plays a crucial role in every modern constitutive theory for soil. In this sense, every such theory is a critical-state theory (see also [74, 91]).

The widespread (and tacit) assumption that the function $e_c = \epsilon_1(p)$ is also valid for non-critical stress states, i.e. for $q \neq mp$, can be questioned. How to assign e_c-values to stress states $q \neq mp$? As mentioned, the relation (Equation 16.9) holds along isochoric (i.e. $\delta = 0$) proportional paths. A generalisation towards proportional paths with $\delta \neq 0$ is:

$$e_c = \epsilon_2(p) = \epsilon_1(p) \cdot (1 + \kappa_2\, \delta). \tag{16.10}$$

As a further generalisation towards non-proportional paths, we introduce now the assertion that e_c is path-dependent according to the following evolution equation, which was found heuristically:

$$\dot{e}_c = \kappa_1 \cdot [\epsilon_2(p) - e_c] \cdot \dot{\varepsilon} + (e_{c0} - e_c)\, \dot{\lambda}/2. \tag{16.11}$$

The first term in Equation 16.11 causes the asymptotic approach $e_c \to \epsilon_2(p)$ with increasing deformation. The second term in Equation 16.11 vanishes for monotonic strain paths $\mathbf{D}^0$ = const, and causes a jump of e_c at jumps of $\mathbf{D}^0$. To illustrate this, consider the following stretching history with a jump at $t = t_0$:

$$\mathbf{D}^0 = \mathbf{D}^{0-} = \text{const} \quad \text{for} \quad t < t_0 \tag{16.12}$$

$$\mathbf{D}^{0-} \to \mathbf{D}^{0+} \quad \text{at} \quad t = t_0 \tag{16.13}$$

$$\mathbf{D}^0 = \mathbf{D}^{0+} = \text{const} \quad \text{for} \quad t > t_0. \tag{16.14}$$

With the jump $\Delta\mathbf{D}^0 := \mathbf{D}^{0+} - \mathbf{D}^{0-}$, the critical void ratio e_c suffers the jump Δe_c at $t = t_0$:

$$\Delta e_c = (e_{c0} - e_c) \int_{-\infty}^{\infty} \frac{\dot{\lambda}}{2} \mathrm{d}t \tag{16.15}$$

$$= (e_{c0} - e_c) \frac{1}{2} |\Delta \mathbf{D}^0| \int_{-\infty}^{\infty} \delta(t - t_0) \mathrm{d}t \tag{16.16}$$

$$= (e_{c0} - e_c) \frac{1}{2} |\Delta \mathbf{D}^0|, \tag{16.17}$$

where $\delta(t - t_0)$ is the Kronecker delta. For a loading reversal $\mathbf{D}^{0+} \to -\mathbf{D}^{0-}$ we have $|\Delta \mathbf{D}^0| = 2$, and this implies that e_c is pulled back to the initial value e_{c0}.

The (numerical) integration (see Section 16.9.2) of the differential equation (16.11) is easy but requires knowledge of the initial value $e_{c0} := e_c(t = 0)$, which may depend on, for example, the method of preparation of the sample, which is known to play an important role in the test results (Fig. 2.13). For the time being, e_{c0} can only be determined by trial and error. The e_{c0} values chosen for the simulations in Sections 16.9.1 and 16.10.1 are shown in the corresponding plots.

16.6 The R-function

The function $\mathbf{R}(\mathbf{D}) = -\exp\left(c_1 \mathbf{D}^{0\star}\right) + c_2 \delta \mathbf{1}$ uses the so-called matrix exponential, which is defined a follows:

$$\exp \mathbf{A} = \exp \begin{pmatrix} A_1 & 0 & 0 \\ 0 & A_2 & 0 \\ 0 & 0 & A_3 \end{pmatrix} = \begin{pmatrix} \exp A_1 & 0 & 0 \\ 0 & \exp A_2 & 0 \\ 0 & 0 & \exp A_3 \end{pmatrix}, \tag{16.18}$$

with A_1, A_2, A_3 being the principal values of $\mathbf{A}$. In barodesy, the argument of the matrix exponential is $c_1 \mathbf{D}^{0\star}$, where $\mathbf{D}^{0\star}$ is the deviatoric part of the unit stretching $\mathbf{D}^0$. Note that the matrix exponential in the expression for $\mathbf{R}(\mathbf{D})$ is also defined for non-rectlinear extensions. The justification for the matrix exponential can be easily explained. For isochoric stretchings (i.e. $\mathbf{D}^\star = \mathbf{D}, \delta = 0, D_1 + D_2 + D_3 = 0$) we require that the corresponding proportional stress path be in the compressive octant of the σ_1-σ_2-σ_3-space, i.e. we require $\sigma_1 \cdot \sigma_2 \cdot \sigma_3 > 0$. The exponential function fulfils this requirement, since it transforms a sum into a product: $\exp(D_1 + D_2 + D_3) = \exp D_1 \cdot \exp D_2 \cdot \exp D_3$.

16.6.1 Deviatoric Implications

Equation 16.7, the 'R-function', maps all volume-reducing ($\mathrm{tr}\mathbf{D} < 0$) proportional strain paths into proportional stress paths within a cone in the stress space with apex at $\mathbf{T} = \mathbf{0}$. Its boundary corresponds to paths with $\mathrm{tr}\mathbf{D} = 0$ and is thus the so-called critical state surface. The equation of this surface can be derived from Equation 16.7 as follows: For isochoric deformations ($\mathrm{tr}\mathbf{D}^0 = 0$) we can eliminate $\mathbf{D}^0$ from (16.7) and obtain:

$$\mathbf{D}^0 = \frac{1}{c_1} \ln(-\mathbf{R}), \tag{16.19}$$

$\ln(-R_1R_2R_3) = 0$. With $\sigma_i^0 = -R_i^0$ the additional requirement $|\mathbf{D}^0| = 1$ leads to:

$$\left(\ln\frac{\sigma_1}{\sqrt[3]{\sigma_1\sigma_2\sigma_3}}\right)^2 + \left(\ln\frac{\sigma_2}{\sqrt[3]{\sigma_1\sigma_2\sigma_3}}\right)^2 + \left(\ln\frac{\sigma_3}{\sqrt[3]{\sigma_1\sigma_2\sigma_3}}\right)^2 = c_1^2. \tag{16.20}$$

Equation 16.20 is homogeneous in the zeroth degree in $\mathbf{T}$ and describes thus a conical surface in the stress space with apex at $\mathbf{T} = \mathbf{0}$. Its intersection with a plane $\mathrm{tr}\mathbf{T} = \mathrm{const}$ is shown in Fig. 16.2. Equation 16.20 can be simplified to:

$$\left(\ln\frac{\sigma_1}{\sigma_2}\right)^2 + \left(\ln\frac{\sigma_1}{\sigma_3}\right)^2 + \left(\ln\frac{\sigma_2}{\sigma_3}\right)^2 = 3c_1^2 \tag{16.21}$$

Fig. 16.2 can be numerically obtained with Equation 16.7 as the locus of stationary stress states that correspond to isochoric stretchings $\mathbf{D}^0, \mathrm{tr}\mathbf{D}^0 = 0$. For any such stretching, Equation 16.7 yields a stress ray $\mathbf{T} = \mu\mathbf{R}, \mu > 0$. Its intersection with a deviatoric plane is a point on the shown curve. Its shape practically coincides [26] with the one given by Matsuoka and Nakai.

16.6.2 Volumetric Implications

Proportional stress paths pertaining to isochoric proportional strain paths define the cone of critical states, whose intersection with a deviatoric plane, as shown in Fig. 16.2, complies well with existing evidence in soil mechanics. However, how about *volume-reducing* proportional strain paths? Does Equation 16.7 yield realistic directions of the corresponding proportional stress paths? In order to verify the proposed Equation 16.7, we resort to experimental evidence by Goldscheider [31], who

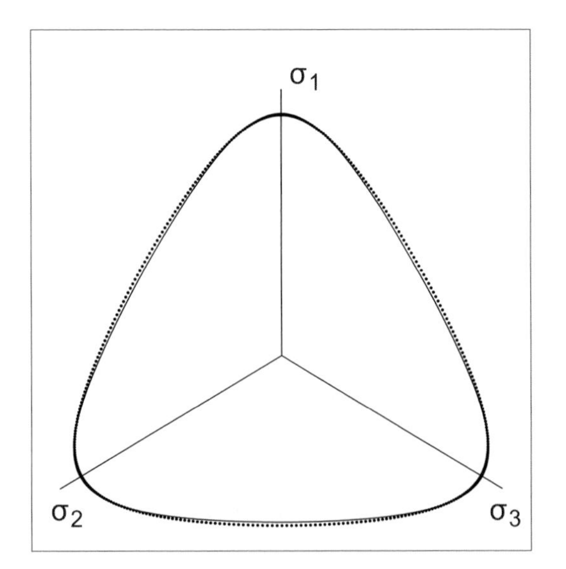

Figure 16.2 Cross section of the critical state surface with a deviatoric plane, obtained with Equation 16.7 (solid line). The dotted line is obtained with the approximation (16.25). It almost coincides with the solid line.

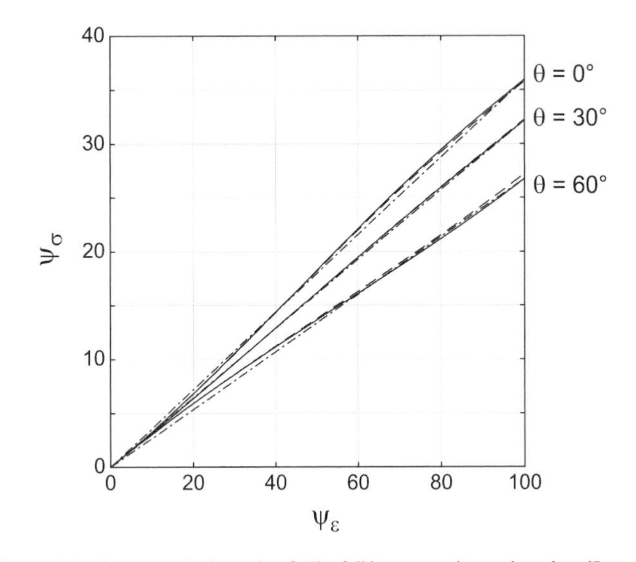

Figure 16.3 ψ_σ in dependence of ψ_ε for various Lode angles θ. The full lines according to barodesy (Equation 16.7), the dashed lines calculated with (Equation 16.25) and the dashdotted lines according to (Equation 16.22) almost coincide.

established (by means of experimental results with sand obtained with a true triaxial apparatus) a relation between the angles ψ_σ and ψ_ε. ψ denotes the angular deviation of a proportional stress or strain path from the principal diagonal of the stress or strain space, i.e. from the lines $\sigma_1 = \sigma_2 = \sigma_3$ and $\varepsilon_1 = \varepsilon_2 = \varepsilon_3$, respectively. Goldscheider obtained the ψ_σ–ψ_ε relation (so-called volumetric flow rule) for the Lode angle $\theta = 0$ and assumed that it holds also for other θ values. This relation reads:

$$\psi_\sigma = \frac{\psi_{\sigma p}}{\psi_{\varepsilon p}} \psi_\varepsilon, \tag{16.22}$$

where the values $\psi_{\sigma p}$ and $\psi_{\varepsilon p}$ ($\approx 100°$) refer to peak states. The simulations with Equation 16.7 and also with the approximation (16.25) are shown in Fig. 16.3. Obviously, the simulations almost coincide with (16.22).

To calculate the relation shown in Fig. 16.3, one needs the correspondence between a unit tensor $\mathbf{A}$, with principal values A_1, A_2, A_3, and the angle ψ_A. This can be obtained with the Haigh–Westergaard variables ξ, ρ, θ:

$$\xi = \frac{1}{\sqrt{3}}\mathrm{tr}\mathbf{A} = \cos\psi_A, \quad \rho = \sqrt{\mathrm{tr}\mathbf{A}^{\star 2}} = \sin\psi_A, \quad \cos(3\theta) = 3\sqrt{6}\frac{\det\mathbf{A}^\star}{(\mathrm{tr}\mathbf{A}^{\star 2})^{3/2}}, \tag{16.23}$$

$$\begin{pmatrix} A_1 \\ A_2 \\ A_3 \end{pmatrix} = \frac{1}{\sqrt{3}}\begin{pmatrix} \xi \\ \xi \\ \xi \end{pmatrix} + \sqrt{\frac{2}{3}}\rho\begin{pmatrix} \cos\theta \\ \cos(\theta - 2\pi/3) \\ \cos(\theta + 2\pi/3) \end{pmatrix}. \tag{16.24}$$

16.6.3 The Matrix Exponential

The matrix exponential (Equation 16.7) is obtained by the exponentials of the principal values of its argument. Note that the derivative of the matrix exponential,

$de^{\mathbf{X}}/d\mathbf{X}$, is extremely complicated, a surprising fact, to be contrasted to the simple derivative of the scalar exponential e^x. Interestingly, a broken Taylor series of the matrix exponential is a good approximation (Figs. 16.2 and 16.3):

$$\exp(a\mathbf{X}) \approx \mathbf{1} + a\mathbf{X} + \frac{1}{2}(a\mathbf{X})^2 + \frac{1}{6}(a\mathbf{X})^3. \tag{16.25}$$

Using this approximation, we can obtain analytically the stiffness matrix as $\partial\mathbf{h}/\partial\mathbf{D}$ which (with rate independence) equals $\partial\mathbf{T}/\partial\mathbf{E}$, with $\mathbf{E}$ being the infinitesimal deformation tensor.

16.6.4 Links to Established Concepts

The following notions from traditional soil mechanics are contained in barodesy. To show this, we consider Equation 16.6 for (either peak or residual) limit states: vanishing stiffness implies $\dot{\mathbf{T}} = \mathbf{0}$. We then have $\mathbf{R}^0 = \mathbf{T}^0$ and $f + g = 0$.

Flow rule: The equation $\mathbf{R}^0(\mathbf{D}^0) = \mathbf{T}^0$ relates the direction of stretching $\mathbf{D}^0$ with a limit stress state $\mathbf{T}^0$.

Friction angle: The friction angle does not explicitly appear in barodesy. Therefore, it does not easily become clear that shear strength increases with normal stress. However, this is in fact the case. For critical limit states ($\delta = 0$), $\mathbf{R}^0$ obtains, according to Equation 16.7, particular values and so does $\mathbf{T}^0$ ($= \mathbf{R}^0$, in this case). The equation $\mathbf{T}^0 =$ const, being homogeneous in $\mathbf{T}$, defines a cone in stress space with apex at $\mathbf{T} = \mathbf{0}$ and this implies that residual (or critical) strength increases with normal stress.

Stress dilatancy: As mentioned, the equality $\mathbf{T}^0 = \mathbf{R}^0$ holds also for peak states. This implies that the aperture angle of the related cone in stress space (and thus the peak friction angle φ_{peak}) is related to the dilatancy at peak, δ_{peak}, a fact which corresponds to the well-known stress-dilatancy relation of Taylor [103], see Equation 9.4. Clearly, the peak corresponds to maximum dilatancy, i.e. to the inflexion point of the volumetric curve ε_v vs. ε_1, obtained from drained triaxial tests.

The equation $f+g = 0$ implies with (16.8) $e_c - e_{peak} + 2c_3\delta_{peak} = 0$, i.e. decreasing e_{peak} increases the dilatancy δ_{peak}, in accordance with the fact that the denser a soil is, the bigger its dilatancy and its friction angle are. In particular, $\delta_{peak} = 0$ for $e_{peak} = e_c$.

16.7 Calibration

In barodesy, only c_1 can be directly determined, and this occurs on the basis of the critical friction angle φ_c [52]:

$$c_1 = \sqrt{\frac{2}{3}} \ln\left(\frac{1 - \sin\varphi_c}{1 + \sin\varphi_c}\right). \tag{16.26}$$

Table 16.1. Material constants for Hostun sand and 'Karslruhe clay'

	Sand	Clay
c_1	−1.025	−0.74
c_2	0.50	0.50
c_3	−0.20	−0.25
c_4	5000	280
c_5	0.50	0.67
κ_1	30	20
κ_2	0	0.24
k_1	1.00	2.16
k_2	0.02	0.24

All other material constants, i.e. c_2, c_3, c_4, c_5, k_1, k_2, κ_1, κ_2, as well as e_{c0}, cannot be directly determined, as they are interrelated in a complex way, and therefore, they have to be determined by trial and error to yield reasonable simulations. Of course, this is a tedious procedure and is not satisfactory. It is hoped that in future some simple approximations will be available. The two sets of material constants given in Table 16.1 are obtained by trial and error in order to provide good simulations and they can serve in future as starting points of searches to adapt the values to other soils. They yield good fits; however, the fact that other values may give even better fits cannot be excluded.

e_{c0} depends also on the sample preparation method and thus it cannot be uniquely attributed to a particular soil.

The material constants, except for c_4, are dimensionless. The dimension of c_4 is Ψ^{1-c_5}, where Ψ is the dimension used for stress. If Ψ = kN/m^2, then $[c_4]$ = (kN/m^2)$^{1-c_5}$. In Equation 16.9 p has to be taken in kN/m^2, otherwise the constant k_1 should be appropriately modified.

16.8 Simulation of Element Tests

The quality of a constitutive relation can be shown by the realistic simulation of element tests. In the ideal case, a constitutive relation will yield realistic simulations of every available test result. In reality, however, experimental outcomes are burdened by stochastic and several types of systematic errors (inhomogeneous deformation, creep, varying sample preparation method, varying types of soil, etc.). Even the standard experiments of soil mechanics exhibit a large scatter, as round-robin tests indicate. At that, constitutive models are never perfect. So, a particular constitutive relation will not precisely simulate all available test results. We should rather ask whether it is capable of reproducing a wide range of behaviour patterns.

16.9 Barodesy for Sand

The material constants listed in Table 16.1 are obtained by trial and error based of experiments with Hostun sand [16].

16.9.1 Simulations of Element Tests with Sand

Figs. 16.4, 16.5, 16.11, 16.12–16.16 and 16.17 show simulations of several types of element tests and show the overall capabilities of barodesy without reference to particular tests. Figs. 16.6–16.10 show simulations contrasted with particular test results taken from [16].

Following the usage in soil mechanics, compressive stresses and strains are taken as positive in the plots (except for volumetric strain). All stresses are given in kPa (or kN/m^2).

The particular results are commented as follows:

Drained triaxial tests: Fig. 16.5 shows the simulation of triaxial tests on a loose ($e_{initial}$ = 0.85) and a dense ($e_{initial}$ = 0.70) sample. The triaxial compressions are followed by triaxial extensions, a procedure which is technically difficult to apply in the laboratory. The influence of lateral stress σ_2 (so-called barotropy) is shown in Fig. 16.17: increasing σ_2 decreases dilatancy and peak friction angle φ_p.

Undrained triaxial tests: Typically, the so-called CU tests consist of two deformation paths with constant stretching **D** each: first a consolidation (volume reduction) and subsequently a shear with volume-preserving **D**. The corresponding stress paths are first a proportional stress path emanating from **T** = **0** and

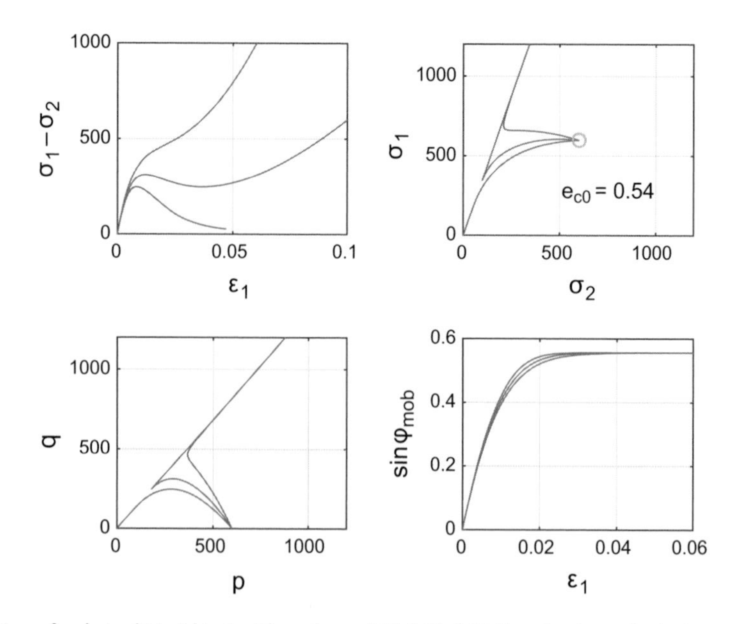

Figure 16.4 Simulations of undrained triaxial tests with sand, e = 0.70, 0.85, 1.00. Note the sharp edge in the stress paths, usually interpreted as phase transformation.

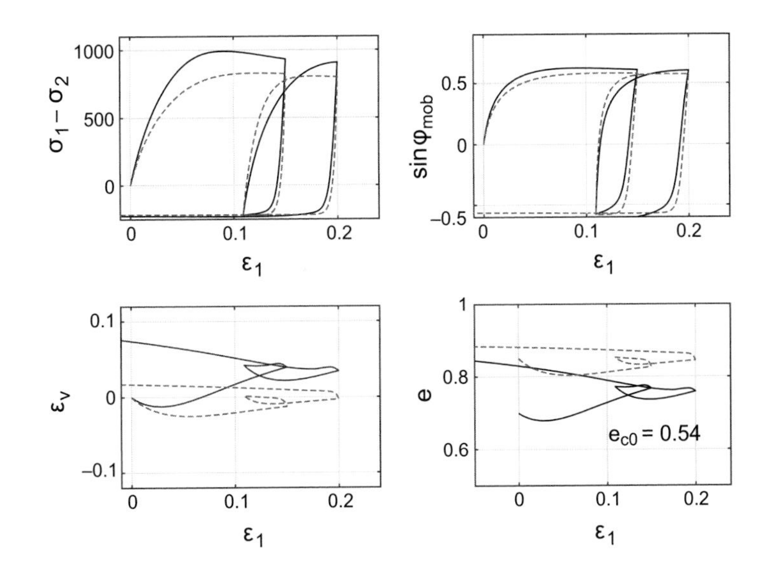

Figure 16.5 Simulations of drained triaxial tests ($\sigma_2 = 300$ kPa) of a loose (dashed) and a dense sample.

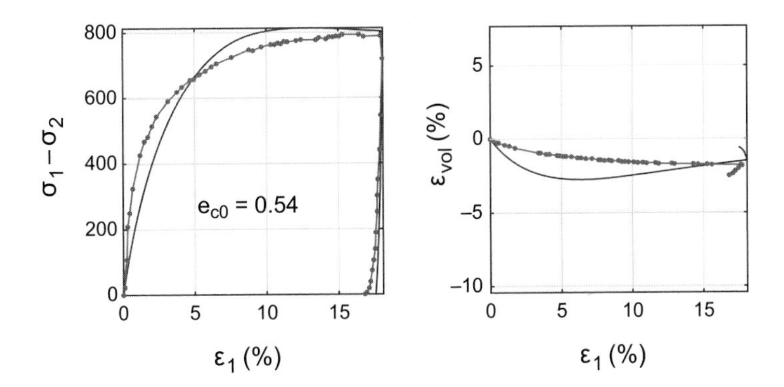

Figure 16.6 Simulation of the triaxial test *host-l-triaxc-cd-300*, $e_{initial} = 0.869$, $\sigma_2 = 300$ kPa.

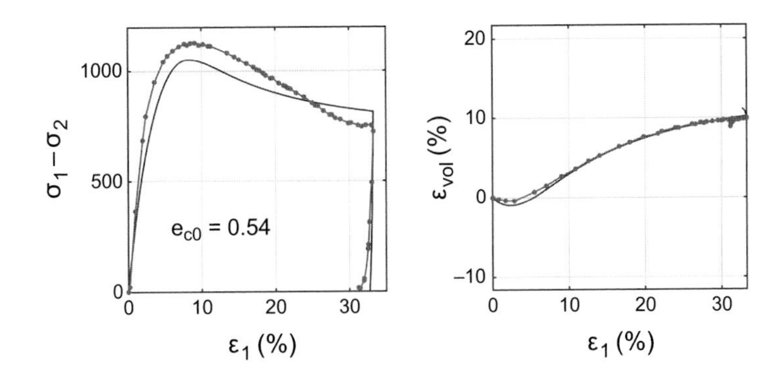

Figure 16.7 Simulation of the triaxial test *host-d-triaxc-cd-300*, $e_{initial} = 0.656$, $\sigma_2 = 300$ kPa.

subsequently a curved path that approaches the CSL and heads either to $\mathbf{T} = \mathbf{0}$ or to $|\mathbf{T}| \to \infty$ (Fig. 16.1). Fig. 16.4 shows the simulations of tests with various void ratios.

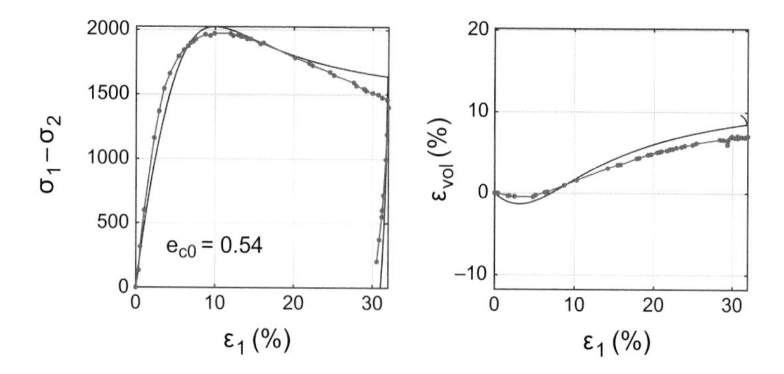

Figure 16.8 Simulation of the triaxial test *host-d-triaxc-cd-600*, $e_{initial} = 0.671$, $\sigma_2 = 600$ kPa.

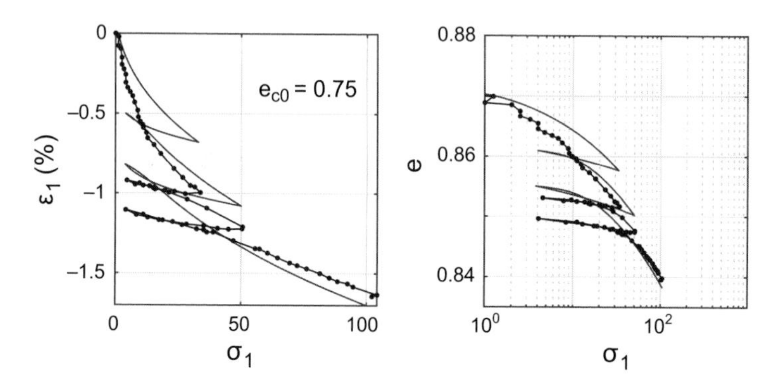

Figure 16.9 Simulation of the oedometric test *host-l-oed*. $e_{initial} = 0.870$ (loose).

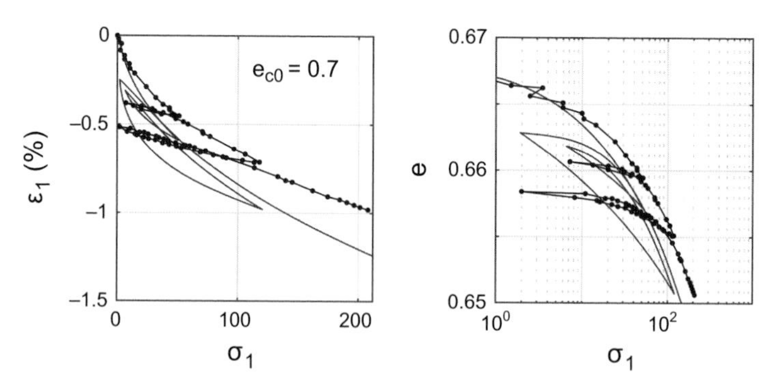

Figure 16.10 Simulation of the oedometric test with sand, *host-d-oed*, $e_{initial} = 0.667$ (dense).

Cyclic undrained tests are important for liquefaction studies. Figs. 16.12 and 16.13 show simulations with cyclic stress and cyclic strain, respectively. The typical ‘butterfly’ shapes appear.

Oedometric tests: Simulations and observed values are shown in Fig. 16.9 for a loose sample and in Fig. 16.10 for a dense one. The stress path at oedometric stress cycles is shown in Fig. 16.11. Upon unloading, the lateral stress σ_2 is less decreased than the axial one.

Simple shear tests: Fig. 16.14 (constant volume) and Fig. 16.15 (constant normal stress) show the simulations. Simple shear tests in the laboratory are rare because

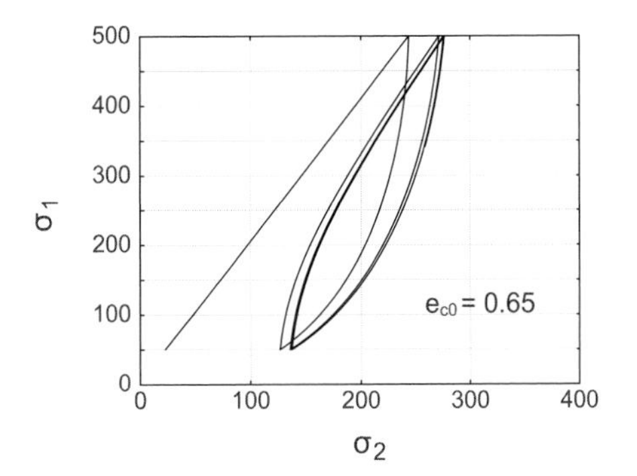

Figure 16.11 Simulated stress path at oedometric cycles, $e_{initial} = 0.65$.

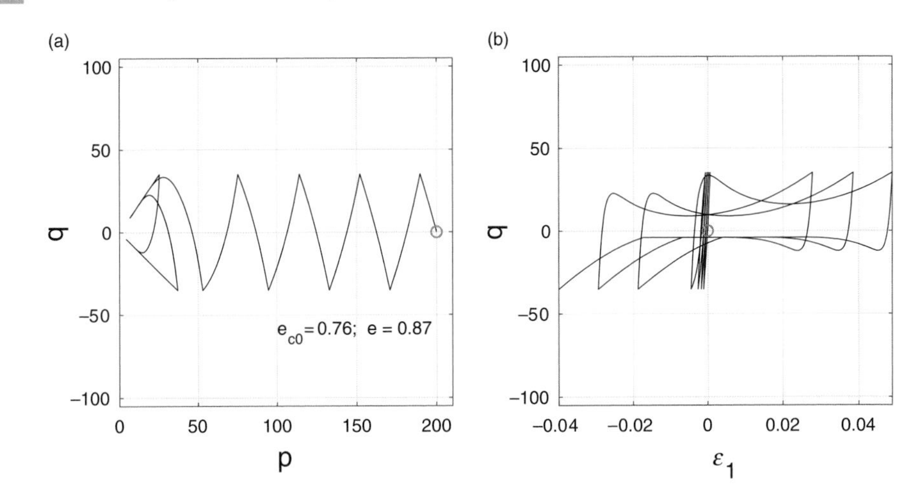

Figure 16.12 Simulation of cyclic stress in undrained triaxial tests with sand. (a) Stress path and (b) stress–strain curve. The amplitude of strain increases gradually (cyclic mobility). The stress path exhibits the typical butterfly pattern.

of the inherent inhomogeneous deformation [10]. The simulated tests are cyclic in strain, and this leads eventually to stationary cycles (also called 'shakedown').

Fig. 16.16 shows simulations of simple shear with stress cycles. As with experiments, the simulated curves tend to stationary (or limit) cycles that are exhibited in the stress path and in the stress–strain curve.

16.9.2 Barodesy in MATLAB

The rates of stress $\dot{\mathbf{T}}$ and critical void ratio $\dot{e}_c$ (written as `Tp` and `ecp` in MATLAB) can be calculated with the following subroutine (constants for Hostun sand):

```
function [Tp,ecp] = stg(T,D,e,ec)
 c=[-1.025 .5 -0.2 5000 .5]; p=-trace(T)/3;E=1.0-0.02*log(p); ecp=30*(E-ec);
 nT=norm(T,'fro'); nD=norm(D,'fro');T0=T/nT; D0=D/nD; delta=trace(D0);
 D0S=D0-delta/3*eye(3); R=-expm(c(1)*D0S)+c(2)*delta*eye(3);
 R0=R/norm(R,'fro');  h=c(4)*nT^c(5); f=ec+c(3)*delta; g=-e+c(3)*delta;
 Tp = h*(f*R0+g*T0)*nD ;
end
```

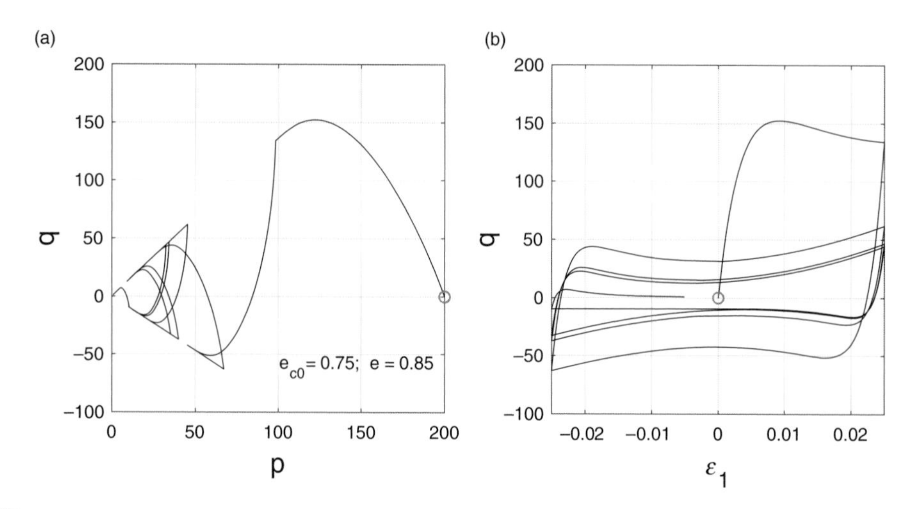

Figure 16.13 Simulations of cyclic strain in undrained triaxial tests with sand. (a) Stress path and (b) stress–strain curve. The stress path exhibits the typical butterfly pattern.

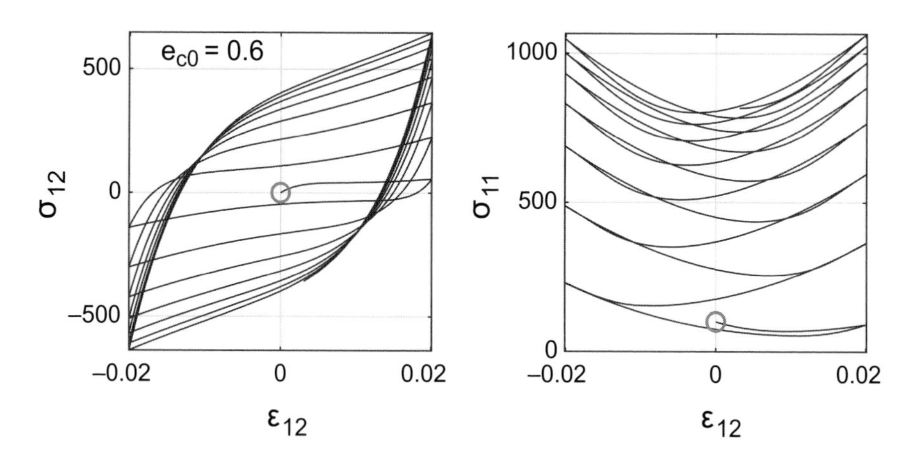

Figure 16.14 Simulation of undrained ($e = 0.70$) simple shear, cyclic strain. Note the gradual onset of stationary cycles (shakedown).

Stress, void ratio and critical void ratio can then be updated in the main program with Euler forward integration:

$$\mathbf{T}(t + \Delta t) = \mathbf{T}(t) + \dot{\mathbf{T}}(t) \cdot \Delta t, \tag{16.27}$$

$$e(t + \Delta t) = e(t) + (1 + e(t)) \cdot \mathrm{tr}\mathbf{D} \cdot \Delta t, \tag{16.28}$$

$$e_c(t + \Delta t) = e_c(t) + \dot{e}_c(t) \cdot \Delta t, \tag{16.29}$$

which yields reasonable results if Δt is sufficiently small. The second term of the right-hand side of Equation 16.11 can be taken into account:

```
nD=norm(D,'fro');D0=D/nD;L=D0-D0alt;  lambda=norm(L,'fro');
D0alt=D0; ec=ec+ecp*dt+lambda*(ec0-ec)/2;
```

The brevity of the subroutine stated above illustrates the simplicity of barodesy.

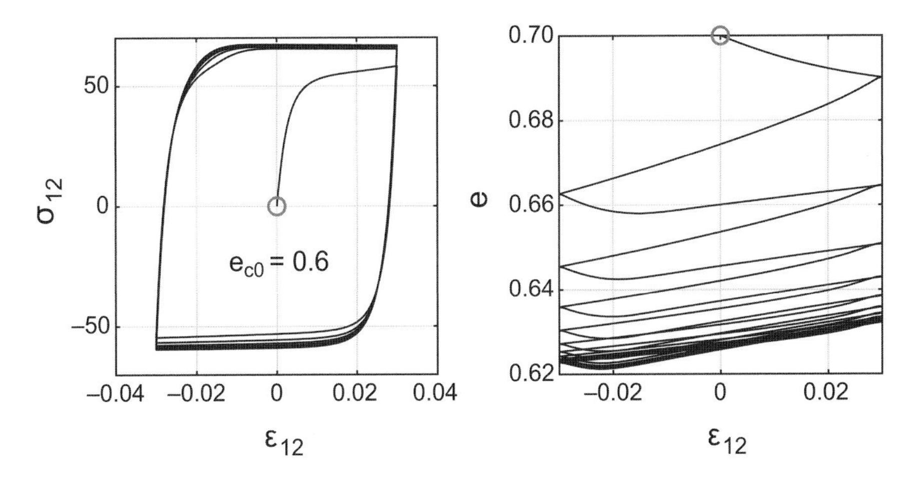

Figure 16.15 Simulation of simple shear with constant normal stress, cyclic strain and initial void ratio: $e = 0.70$. Note the gradual onset of stationary cycles (shakedown).

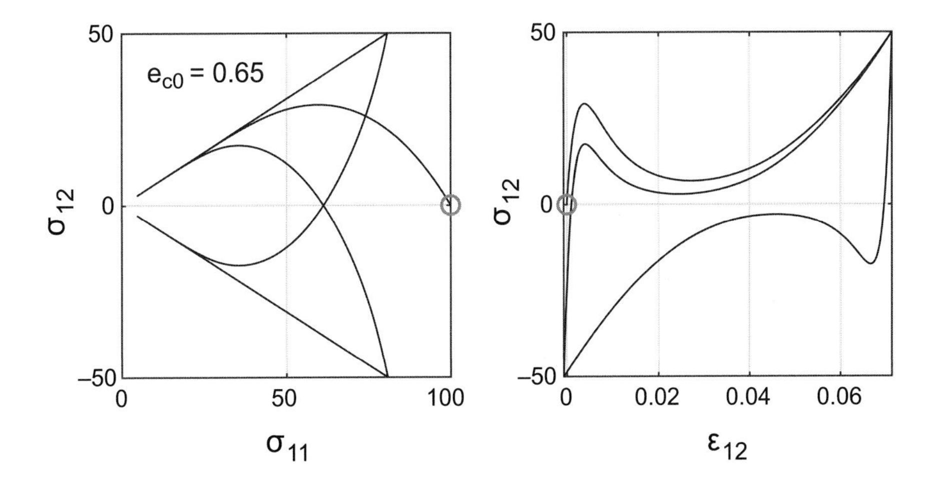

Figure 16.16 Simulation of undrained simple shear with sand, cyclic stress, $e = 0.70$. Stationary (or limit) cycles (shakedown) set in immediately. The stress path exhibits the typical butterfly pattern.

16.10 Barodesy for Clay

Sand and clay are the two main types of soil, and constitutive relations are usually formulated either for sand or for clay. Barodesy, as presented here, applies also to clay with the same expressions as the ones used for sand without any reference to concepts of plasticity theory.

The material constants listed in Table 16.1 are obtained by trial and error based on experiments by Wichtmann with a Kaolin silt called 'Karlsruhe clay' [115]. The results of these experiments are typical for clay.

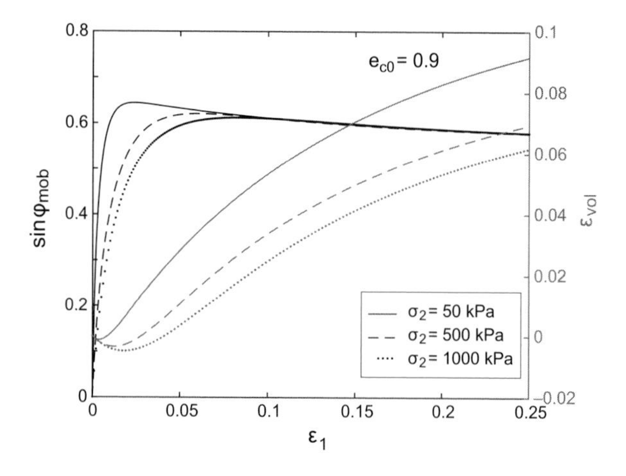

Figure 16.17 Barotropy: Simulations of drained triaxial compressions of sand at lateral stresses from 50 to 1000 kPa, simulated with barodesy. $e_{initial} = 0.70$.

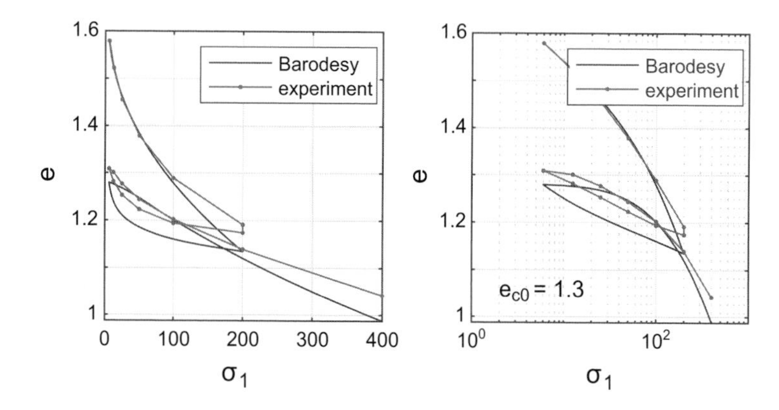

Figure 16.18 Simulation of an oedometric test with 'Karlsruhe clay' vs measured values. Data taken from Wichtmann [115].

16.10.1 Simulations of Element Tests with Clay

Oedometric test: The simulation is shown in Fig. 16.18 together with experimental results taken from Wichtmann [115]. Un- and reloading are rather satisfactorily simulated.

Drained triaxial tests: Barodesy predicts that normally consolidated clay contracts when sheared (Fig. 16.20), whereas overconsolidated clay expands (Fig. 16.19), see [36, 80, 100]. The dilatancy of overconsolidated clay is important in practice, as it provides temporarily increased strength (cohesion) to dense clay due to the suck in of water and the related long-lasting negative pore pressure (suction). The duration of suction is shortened by the generation of thin dilating shear bands ('internal drainage'). This occurs also in true (Hambly type) triaxial/biaxial deformation of clay samples in the laboratory, see Fig. 17.5 as well as Figs. 5.25 in [62] and 4.5 in [107].

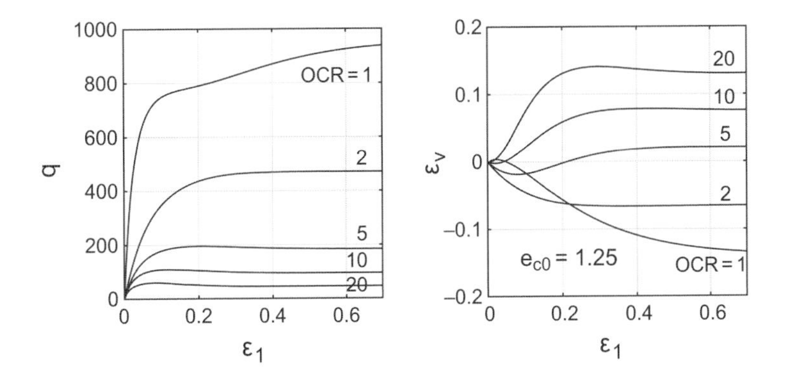

Figure 16.19 Simulated q-ε_1 and ε_v-ε_1 curves of p = const drained triaxial tests with clay. All samples have been consolidated to $p_c = 1000$ kN/m^2 and then unloaded, so that $OCR = 1, 2, 5, 10, 20$. Compare with Fig. 2.6.

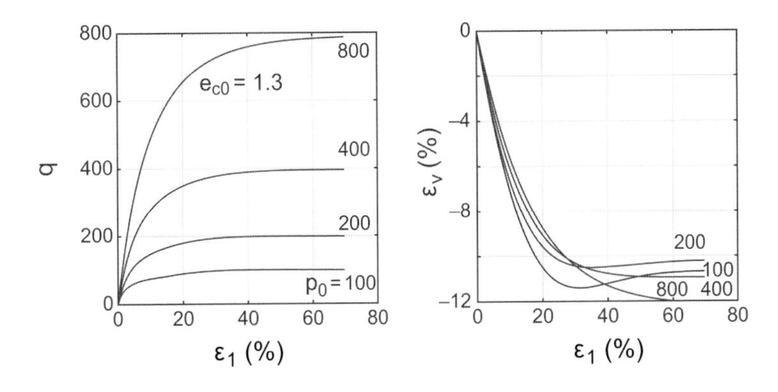

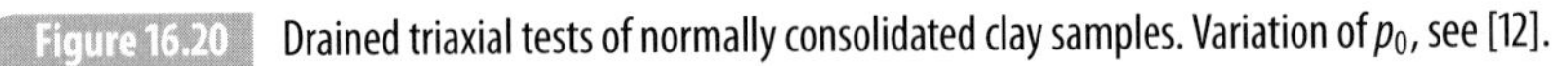
Figure 16.20 Drained triaxial tests of normally consolidated clay samples. Variation of p_0, see [12].

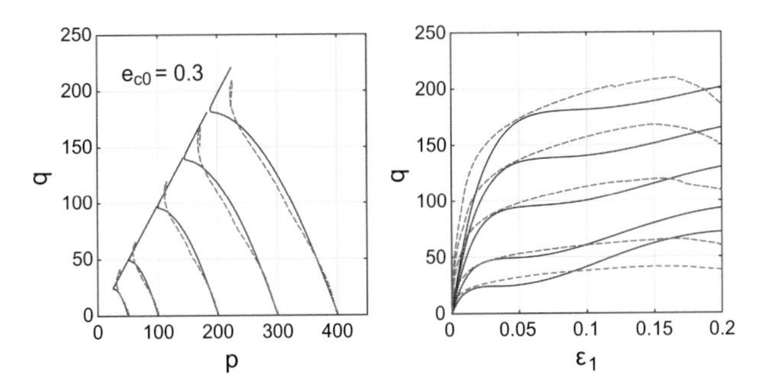

Figure 16.21 Undrained triaxial tests of normally consolidated clay. Variation of p_0. Dashed: experimental results taken from Wichtmann [115].

Undrained triaxial tests: Fig. 16.21 shows simulations of undrained triaxial tests with normally consolidated samples with various initial values p_0. Fig. 16.22 shows simulations of undrained triaxial tests with various *OCR*-values (*OCR* is the overconsolidation ratio).

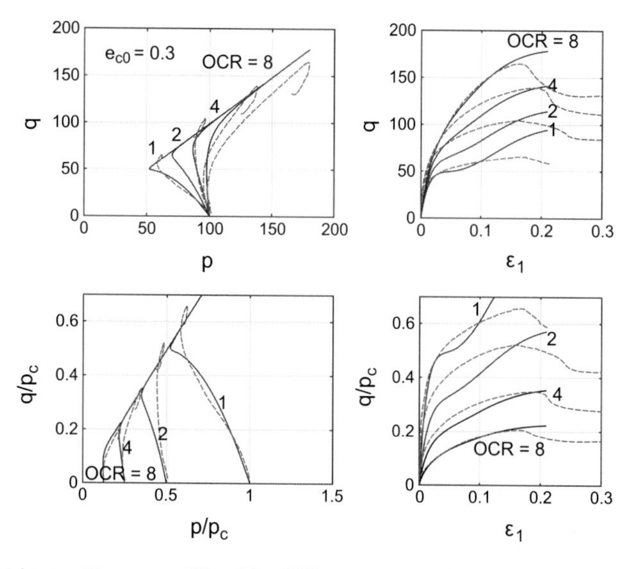

Figure 16.22 Undrained triaxial tests with overconsolidated clay. $OCR = 1, 2, 4, 8$, with the same consolidation pressure p_c. Dashed: Experimental results taken from [115].

16.11 Reflecting upon Barodesy

Looking back to several decades of research and to the many different proposed constitutive relations one may ask: Do we still need a convincing constitutive relation for soil?

The behaviour of a material (or indeed of a virus, a consumer, a politician, or an entire society) can never accurately be described, or predicted, however. So, a behavioural theory is only approximate and can always be improved. Can it only be improved through increasing complexity? Or should we expect such a theory to remain intelligible? For those who consider constitutive theories as black boxes, hidden in computers, these questions appear irrelevant. Nevertheless, intelligible theories are indispensable for planning and evaluating laboratory tests and, more importantly, for understanding the behaviour of soil.

Researchers accustomed to current constitutive relations, especially the ones based on the theory of plasticity, might be reluctant to consider a novel one, however simple it is. This is easily explained: confusing complexity undermines mastering and thus encourages the urge to conform to the mainstream. At any rate, complexity is not a (necessary) attribute of physical laws.

The theory published here, for all its simplicity, is not trivial and reflects the author's meticulous track record in the field since 1977 [53]. His aim in developing barodesy (as well as hypoplasticity, its precursor) was to describe the mechanical behaviour of soil by means of a unique tensor-valued evolution equation according to rational mechanics, and to avoid thus loading–unloading distinctions based on yield surfaces, as postulated by plasticity theory.

16.12 Numerical Simulation of Element Tests

Constitutive equations can be used in numerical codes to solve initial boundary value problems (IBVPs). This is often done via a subroutine, such as UMAT for ABAQUS. For the simulation of element tests, constitutive equations can be integrated numerically, no finite element (FEM) code is needed.

Constitutive equations of the form $\dot{\mathbf{T}} = \mathbf{h}(\mathbf{T}, \mathbf{D}, e)$ are ordinary differential equations of first order. Time integration yields the stress path and the σ-ε-curves. We should distinguish between kinematic and static boundary conditions (BCs):

Kinematic BC: (e.g. oedometer or undrained triaxial test) $\mathbf{D}$ is given, the constitutive law provides the corresponding stress rate $\dot{\mathbf{T}}$. The prerequisite is that $\mathbf{T}(t = 0)$ and $e(t = 0)$ are known. For a small time step Δt one then obtains the stress and strain increments $\Delta\mathbf{T} = \dot{\mathbf{T}}\Delta t$ and $\Delta\mathbf{E} = \mathbf{D}\Delta t$. For sufficiently small Δt this time integration converges to the real solution ('Euler forward').

Static BC: Not all components of the stretching tensor $\mathbf{D}$ are given. For example, in the conventional triaxial test the lateral stretching $D_2(= D_3)$ has to be determined in each step so as to provide $\dot{\sigma}_2 = 0$. This is achieved by solving the algebraic equation $\dot{\sigma}_2 = 0 \leadsto D_2 = \ldots$ i.e. one has to determine the unknown components of $\mathbf{D}$ numerically at each time step by inversion of the constitutive law. This is always possible as long as $\mathrm{tr}(\dot{\mathbf{T}}\mathbf{D}) > 0$ holds. Subsequently, one proceeds as in the case of the kinematic BC.

The MATLAB codes for oedometric and conventional triaxial tests read as follows:

```
function oedo % Simulation of oedometer test with loading and unloading
close all;clear all;
ec0 = 0.60; ec=ec0; e = 0.80;
D = [-1 0 0; 0 0 0; 0 0 0];
c1 = -1.025; c2 = 0.45; R=-expm(c1*D)-c2*eye(3);
K0 = R(2,2)/R(1,1); T1A = -50.;
T = [T1A 0 0; 0 K0*T1A 0; 0  0  K0*T1A];
dt = 0.0001;
T1(1)=-T(1,1);T2(1)=-T(2,2); eps1(1)=0;
T1upper=500; T1lower=50;
loading=1; i=1;
while (-T(1,1)<T1upper & loading>0)
     [Tp,ecp] = stg(T,D,e,ec);
     ec=ec+ecp*dt; e=e+(1+e)*trace(D)*dt;
     i=i+1;
     eps1(i)=eps1(i-1)-D(1,1)*dt ;
     T=T+Tp*dt ;
     T1(i)=-T(1,1); T2(i)=-T(2,2);
end
loading=-1;D=-D;
while (-T(1,1)>T1lower)
   [Tp,ecp] = stg(T,D,e,ec);
   ec=ec+ecp*dt; e=e+(1+e)*trace(D)*dt;
   i=i+1;
   eps1(i)=eps1(i-1)-D(1,1)*dt ;
   T=T+Tp*dt;
   T1(i)=-T(1,1); T2(i)=-T(2,2);
end
```

continued on the next page

```
subplot(1,2,1)
plot(T1,eps1,'-k')
xlabel('\sigma_{11} (kN/m^2)','FontSize',15)
ylabel('\epsilon_{11}','FontSize',15)
set(gca,'YDir','reverse')

subplot(1,2,2)
plot(T2,T1,'-k'); hold on
xlabel('\sigma_{2} ')
ylabel('\sigma_{1} ')
axis image
```

```
function triax   % triaxial loading and unloading
close all; clear all
ec0 = 0.54;ec = ec0;e0 = 0.66; e=e0; T_ini=-300;emax=0.3;
dt = 0.001;
T = [T_ini 0 0; 0 T_ini 0; 0 0 T_ini];
D(1,1) = -1; i=1; eps11(i)=0;evol(i) = 0; Tdev(i) = 0;
while (eps11(i) < emax )
    D2start = - D(1,1);
    options = optimset('Display','off','TolFun',0.0001,
                     'MaxIter',50,'Algorithm','levenberg-marquardt');
    Y = @(D2start) sigma2(D2start,D,T,e,ec);
    [D2start,fval,exitflag,output] = fsolve(Y,D2start,options);
    D=[D(1,1) 0 0 ; 0 D2start 0 ; 0  0 D2start];
    [Tp,ecp] = stg(T,D,e,ec);
    ec=ec+ecp*dt; T=T+Tp*dt; e=e+(1+e)*trace(D)*dt;
    Tdev(i+1)=-(T(1,1)-T(2,2));
    eps11(i+1)=eps11(i)-D(1,1)*dt;
    evol(i+1)=evol(i)+trace(D)*dt;
    i = i+1;
end
D(1,1) = 1;
while (Tdev(i) > 0 )
    D2start = - D(1,1)/2;
     options = optimset('Display','off','TolFun',0.0001,
                    'MaxIter',50,'Algorithm', 'levenberg-marquardt');
      Y = @(D2start) sigma2(D2start,D,T,e,ec);
   [D2start,fval,exitflag,output] = fsolve(Y,D2start,options);
    D = [D(1,1) 0 0 ; 0 D2start 0 ; 0  0 D2start];
    [Tp,ecp] = stg(T,D,e,ec);
    ec = ec + ecp * dt;
    T = T + Tp*dt;b = T(2,2);
    e = e + (1+e)*trace(D) * dt;
    Tdev(i+1) = -(T(1,1)-T(2,2));a = Tdev(i+1);
    eps11(i+1) = eps11(i) - D(1,1)*dt;
    evol(i+1) = evol(i) + trace(D)*dt;
    i = i+1;
end

subplot(1,2,1)
plot(eps11*100,Tdev,'-k');
xlabel(' \epsilon_1 (%)');
ylabel(' \sigma_1-\sigma_2 ');
axis([0  35  0 1200])
```

continued on the next page

```
pbaspect([1 1 1])
grid on

subplot(1,2,2)
plot(eps11*100,evol*100,'-k');
xlabel(' \epsilon_1 (%)');
ylabel(' \epsilon_{vol} (%)');
axis equal
pbaspect([1 1 1])
```

The next code is the algebraic equation to be solved by the subroutine 'fsolve'. It expresses the requirement $\sigma_2 = \text{const}$

```
function fun = sigma2(D2start,D,T,e,ec)
   D = [D(1,1) 0 0 ; 0 D2start 0 ; 0  0 D2start];
  [Tp,ecp] = stg(T,D,e,ec);
  fun =Tp(2,2);
end
```

Exercise 16.12.1 Calculate the value of K_0 for Hostun sand and clay using the material constants of Table 16.1.

Exercise 16.12.2 Calculate the dilatancy δ at the peak of a triaxial compression test with Hostun sand conducted at the lateral stress of $\sigma_2 = 400$ kPa and the initial void ratio of $e = 0.72$.

Exercise 16.12.3 Calculate the value of σ_3 at the peak of a biaxial compression test ($\varepsilon_3 \equiv 0$) with Hostun sand conducted at the lateral stress of $\sigma_2 = 500$ kPa and the initial void ratio of $e = 0.75$.

17 Uniqueness

17.1 Meaning of Uniqueness and Related Notions

With their calculations civil engineers aim to ensure that no collapse will occur. To a lesser degree they also aim to predict deformations. Usually, collapse is conceived as a distinct event that can be mathematically described by an algebraic equation, the so-called limit condition. This approach overlooks the fact that collapse is virtually nothing but exaggerated deformation that cancels the operability of a structure. At any rate, collapse is related with the onset of some unpleasant effects. In dealing with such problems, one faces notions such as:

- *uniqueness* of the solution of the underlying initial boundary value problem (IBVP). The process path encounters a so-called bifurcation when the solution is no more unique.
- *invertibility* of the constitutive equation $\dot{\mathbf{T}} = \mathbf{h}(\mathbf{T}, \mathbf{D}, \ldots)$, i.e. the existence of a unique function $\mathbf{D} = \mathbf{g}(\mathbf{T}, \dot{\mathbf{T}}, \ldots)$.
- *stability*, is given if sufficiently small perturbations do not grow with time.
- *well-posedness*, is a notion introduced by Hadamard. It is given when the solution

  1. exists,
  2. is unique and
  3. is stable.

  Lack of well-posedness is called 'ill-posedness' and is met by 'regularisation', i.e. introduction of additional requirements, which may be rather arbitrary.

Theorems for uniqueness are established mainly for linear elastic materials, i.e. for constitutive laws of the type $\sigma_{ij} = c_{ijkl}\varepsilon_{kl}$, e.g. for Hooke's law with $c_{ijkl} = \lambda\delta_{ij}\delta_{kl} + \mu(\delta_{ik}\delta_{jl} + \delta_{il}\delta_{jk})$. The following properties can be assigned to the stiffness matrix c_{ijkl} [42]:

positive definiteness: $c_{ijkl}\varepsilon_{ij}\varepsilon_{kl} > 0$ for all ε_{ij}
strong ellipticity: $c_{ijkl}\eta_i\eta_k\xi_j\xi_l > 0$ for arbitrary vectors η_i and ξ_i.

17.2 Uniqueness in Element Tests

Consider a constitutive equation of the rate type $\dot{\mathbf{T}} = \mathbf{h}(\mathbf{T}, \mathbf{D})$, which applies equally to elastoplastic, hypoplastic and barodetic constitutive equations. It can also be written in the form $\dot{\mathbf{T}} = \mathbf{HD}$, with $\mathbf{H}(\mathbf{D}) := \partial\mathbf{h}(\mathbf{T}, \mathbf{D})/\partial\mathbf{D}$ being the stiffness matrix.

We investigate the following question: Under which conditions can we uniquely obtain an affine deformation within a sample whose boundary undergoes an affine motion $\mathbf{v}(\mathbf{x}_{boundary}) = \mathbf{A}\mathbf{x}_{boundary}$ with $\mathbf{A}$ = const?

Clearly, the affine motion $\mathbf{v} = \mathbf{A}\mathbf{x}$ is a solution of the boundary value problem, if we neglect gravity. Now we will check whether this solution is unique. Assume that there also exists another solution $\bar{\mathbf{v}} \neq \mathbf{v}$. Denoting differences with the symbol δ, e.g. $\delta\mathbf{v} = \mathbf{v} - \bar{\mathbf{v}}$, we observe that $\delta\mathbf{v}$ vanishes at the boundary. The equilibrium equation reads $\nabla \cdot \mathbf{T} = \mathbf{0}$, and the continued equilibrium reads $\nabla \cdot \dot{\mathbf{T}} = \mathbf{0}$. The same equations also hold for the stress difference $\delta\mathbf{T} := \mathbf{T} - \bar{\mathbf{T}}$: $\nabla \cdot \delta\mathbf{T} = \mathbf{0}$ and $\nabla \cdot \delta\dot{\mathbf{T}} = \mathbf{0}$. Now we consider the integral $I := \int_V \nabla \cdot (\delta\dot{\mathbf{T}}\delta\mathbf{v})\mathrm{d}V$ and apply the divergence (or Gauss) theorem. The resulting integral

$$I = \int_V \nabla \cdot (\delta\dot{\mathbf{T}}\delta\mathbf{v})\,\mathrm{d}V = \int_S \delta\dot{\mathbf{T}}\delta\mathbf{v} \cdot \mathbf{n}\,\mathrm{d}S = 0 \tag{17.1}$$

vanishes, because $\delta\mathbf{v}$ vanishes on the surface S. Vanishing of I implies:

$$I = \int_V \nabla \cdot (\delta\dot{\mathbf{T}}\delta\mathbf{v})\,\mathrm{d}V = \int_V \delta\dot{\mathbf{T}}\,(\nabla \cdot \delta\mathbf{v})\,\mathrm{d}V + \int_V \delta\mathbf{v}\,(\nabla \cdot \delta\dot{\mathbf{T}})\,\mathrm{d}V = 0. \tag{17.2}$$

The second integral on the right-hand side vanishes due to continued equilibrium. Thus, for non-uniqueness must hold:

$$\int_V \delta\dot{\mathbf{T}} \cdot \nabla\delta\mathbf{v}\,\mathrm{d}V \equiv \int_V \delta\dot{\mathbf{T}} \cdot \delta\mathbf{D}\,\mathrm{d}V = 0, \tag{17.3}$$

which is impossible if $\delta\dot{\mathbf{T}} \cdot \delta\mathbf{D} > 0$ holds everywhere (the expression $\delta\dot{\mathbf{T}} \cdot \delta\mathbf{D}$ denotes the same as $\mathrm{tr}(\delta\dot{\mathbf{T}}\delta\mathbf{D})$). Hence, the condition

$$\mathrm{tr}(\delta\dot{\mathbf{T}}\;\delta\mathbf{D}) > 0 \tag{17.4}$$

implies uniqueness [42].

If the relation $\dot{\mathbf{T}} = \mathbf{H}\mathbf{D}$ is linear in $\mathbf{D}$ (i.e. if the stiffness matrix $\mathbf{H}$ does not depend on $\mathbf{D}$), then we can conclude: $\mathrm{tr}(\delta\dot{\mathbf{T}}\delta\mathbf{D}) > 0$ implies $\mathrm{tr}(\dot{\mathbf{T}}\mathbf{D}) > 0$ and vice versa. The quantity $\mathrm{tr}(\dot{\mathbf{T}}\mathbf{D})$ is called 'second-order work' and its positiveness implies, as stated, uniqueness. Positive second-order work means also that the stiffness matrix is positive definite: $c_{ijkl}\xi_{ij}\xi_{kl} > 0$. As long as uniqueness is guaranteed, the deformation of a soil sample can be controlled by prescribing displacements on its boundary. Therefore, this condition is also called 'controllability'. With respect to linear constitutive models, Nova [70] pointed to the fact that controllability guarantees the uniqueness of general loading programmes, i.e. prescriptions of n independent components of $\dot{\varepsilon}_{ij}$ and the remaining $6 - n$ independent components of $\dot{\sigma}_{kl}$, with $n = 0, \ldots, 6$. The response to this loading consists of the corresponding $6 - n$ components of $\dot{\varepsilon}_{ij}$ and the remaining n components of $\dot{\sigma}_{kl}$. The response is unique, if the symmetric part of the stiffness matrix c^s_{ijkl} is positive definite, or equivalently, if the 'second-order work' W_2 is positive: $W_2 := \dot{\sigma}_{kl}\dot{\varepsilon}_{ij} > 0$.

17.3 Shear Bands and Faults

The deformation of granular materials has the tendency to concentrate ('localise') into thin shear bands (Figs. 17.1–17.3). The appearance of shear bands is a sort of

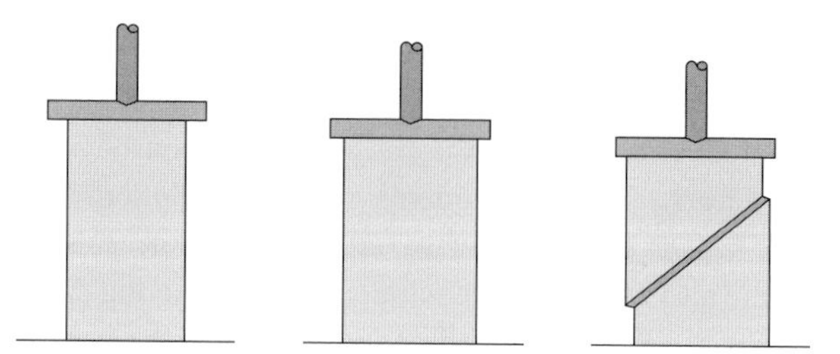

Figure 17.1 Compression of a soil sample in the triaxial apparatus. The initially homogeneous deformation is terminated by the spontaneous appearance of a shear band. Reproduced from [57], courtesy of Springer Nature.

Figure 17.2 Shear bands in nature.

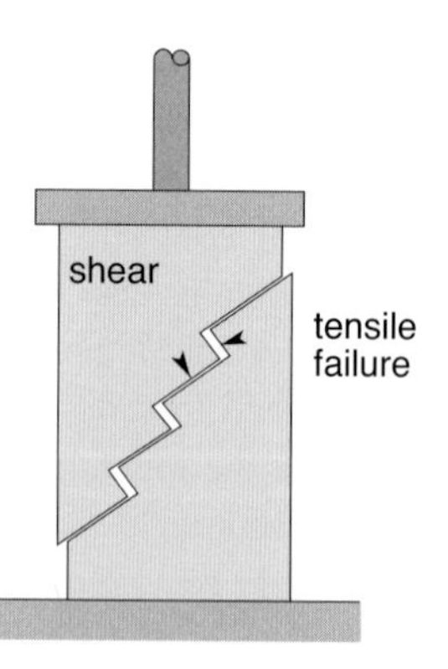

Figure 17.3 A shear band can be combined with tensile cracks. Reproduced from [57], courtesy of Springer Nature.

pattern formation and can be observed through many scales of geometric size, from small soil samples in triaxial tests up to faults in geology (e.g. the San Andreas fault).

It is a challenging question to find under which conditions a shear band can appear, and moreover, to predict its inclination. This is relatively simple using the Mohr–Coulomb failure condition (Fig. 17.4), but here we consider an approach based on the underlying constitutive model. After the appearance of a shear band, the

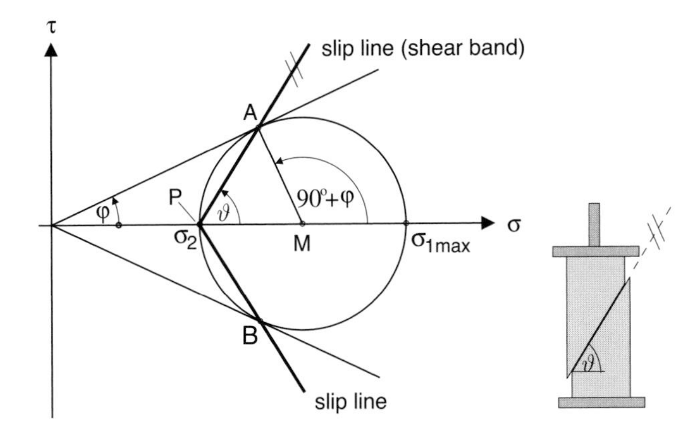

Figure 17.4 Inclination of a shear band in a triaxial test according to Mohr–Coulomb failure criterion. Reproduced from [57], courtesy of Springer Nature.

deformation proceeds as simple shear within the shear band, whereas no deformation occurs outside (Fig. 6.3). The parts of the body separated by the shear band move like rigid bodies that slide relative to each other. Since no deformation takes place in the rigid blocks, the stress there does not change anymore and, consequently, the stress vector $\mathbf{t}$ acting in the interface between shear band and rigid block, remains constant. This can be expressed by the following equation:

$$\dot{\mathbf{t}} = \dot{\mathbf{T}}\mathbf{n} = \mathbf{0}, \tag{17.5}$$

where $\mathbf{n}$ is the unit vector normal to the shear band and $\dot{\mathbf{T}}$ results from the deformation in the shear band according to Equation 6.21. Equation 17.5 governs the appearance of a shear band: once this equation has a solution, a shear band *can* appear, and this signifies that the solution of the IBVP of the deformation of the sample is no more unique. If the stress rate $\dot{\mathbf{T}}$ is expressed using a realistic constitutive equation, $\dot{\mathbf{T}} = \mathbf{h}(\mathbf{T}, \mathbf{D}, e)$, and $\mathbf{D}$ describes simple shear according to Equation 6.21, we can investigate whether this equation has a solution. If this is the case, a shear band *can* appear, and the corresponding $\mathbf{D}$ tensor indicates the inclination of the shear band and the dilatancy occurring therein. Using hypoplasticity, the author presented the first prediction for the spontaneous appearance of a continuous shear band in the course of a biaxial compression test with a sand sample [55, 56, 101]. The formation of a shear band can be instantaneous or gradual. Interestingly, small precursors of a continuous shear band are observed in early stages of the test [19].

As for the geometric compatibility of the deformation of a sample with shear bands, it should be pointed out that the statements of Section 10.4 also hold here. The shear bands divide the sample into individual 'rigid' blocks that slide relative to each other, and the accompanying squeezing of the edges can be disregarded, see Fig. 17.5.

17.3.1 Vortices

In some cases, the deviation velocity field $\mathbf{v}'$ is considered as fluctuation superimposed on the affine velocity field $\bar{\mathbf{v}}$:

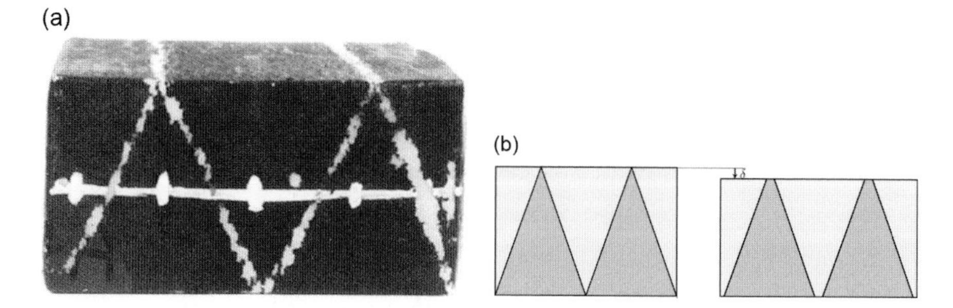

Figure 17.5 (a) Multiple shear bands in a sample compressed in plane deformation. Reproduced from [18] represent a collapse mechanism with rigid blocks sliding relative to each other. (b) The squeezing of the edges is negligible.

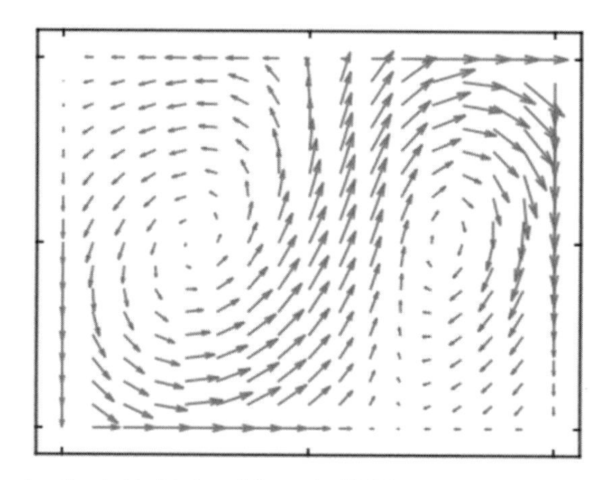

Figure 17.6 Numerically obtained vortices in biaxial plane deformation [5, 58].

$$\mathbf{v} = \bar{\mathbf{v}} + \mathbf{v}'. \tag{17.6}$$

The fluctuation field exhibits characteristic vortex patterns (Fig. 17.6), as observed in experiments [1] and numerical simulations with the material point method [58] and the discrete element method for the simple shear test [105], the direct shear test [59], the triaxial test [76] and the biaxial test [81].

The physical explanation of the vortices is still speculative. Shear motion in a contractant granular body reduces the hydrostatic stress, and hence, reduces the overall stiffness. Thus, the superposition of vortices to an affine deformation implies a superimposed shear, and this reduces the volumetric stiffness, and consequently, also the overall stiffness of contractant media.

18 Symmetry

18.1 General Remarks

The notion of symmetry has interesting and deep-rooted connotations that prove to be very helpful in understanding and describing physical phenomena. To put it somewhat casually: an architect can save half of his work by designing a symmetrical building (Fig. 18.1). Thinking about symmetry means thinking about transformations that leave some properties or relations unchanged, i.e. invariant. Isotropy is one example of symmetry. It is related with the following question: Which rotations leave the mechanical behaviour of a material unchanged? Another issue refers to the units we use to describe a problem. For example, we can measure a length in feet or metres. The considered physical relation should not depend on the (arbitrarily chosen) units. The invariance with respect to units is an important symmetry and serves as a basis of our scaling laws and the definition of mechanical similarity (Section 18.5). Thus, it is the basis of model tests and so-called physical simulation (as contrasted to numerical simulation). Invariance with respect to time scale gives rise to the so-called rate independence.

In general, invariance makes it possible to introduce equivalence classes of similar objects and relations. When two objects are in some sense similar, then they are in a sense equivalent, and it suffices to consider only one of them.

18.2 Principle of Material Frame Indifference

This principle is rather abstract and refers to the proper formulation of constitutive equations. Previous names of this principle are the 'principle of the isotropy of space' and the 'principle of material objectivity'.

Constitutive equations are relations between stress and deformation tensors or of their time rates. When formulating such relations, one should make sure that they are not influenced by the (arbitrary) choice of the frame of reference. The resulting rules [109] are subtle. Examples of non-objective constitutive equations are: $\sigma_{ij} = \text{const} \cdot \varepsilon_{11} \cdot \varepsilon_{ij}$ and $\sigma_{ij} = \text{const} \cdot t \cdot \varepsilon_{ij}$. In the first constitutive equation, the strain component ε_{11} appears, which depends on the arbitrary choice of the coordinate system, in the second constitutive equation the time t appears, which depends on the arbitrary choice of the time zero point.

We consider the motion $\mathbf{x}(\mathbf{X})$. By superimposing a translation $\mathbf{q}$ and a rotation $\mathbf{Q}$, we get the motion:

Figure 18.1 Symmetrical building.

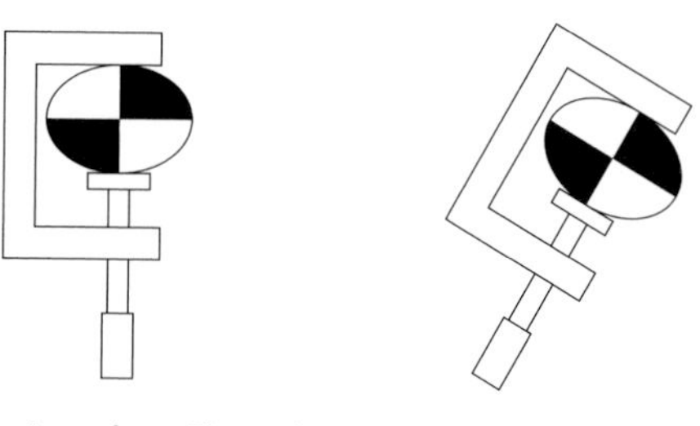

Figure 18.2 A body is deformed by tightening a clamp. The resulting stress states in the body should be the same (and differ only by the resulting rotation) whether or not you rotate the clamp during this process.

$$\mathbf{x}^*(\mathbf{X}, t) = \mathbf{q}(t) + \mathbf{Q}(t)\ [\mathbf{x}(\mathbf{X}, t) - \mathbf{o}]. \tag{18.1}$$

So, $\mathbf{x}$ and $\mathbf{x}^*$ differ by a rigid body motion (Fig. 18.2). They are called *equivalent motions*. Note that there is no way to distinguish whether an object moves or whether our frame of reference moves.

The principle of material frame indifference states that equivalent motions should lead to stresses $\mathbf{T}$ and $\mathbf{T}^*$, respectively, which are linked via the relationship $\mathbf{T}^* = \mathbf{QTQ}^T$ (see Section 18.2.1).

18.2.1 Transformation of Objective Tensors with Rotation of the Frame of Reference

Strictly speaking, a change of the time zero point belongs to a change of the reference system. Leaving this and the trivial case of translation aside, we consider only a rotation $\mathbf{Q}$.

We consider the relationship

$$\mathbf{y} = \mathbf{Ax} \tag{18.2}$$

and rotate the reference frame by $-\mathbf{Q}$, hence $\mathbf{y}^* = \mathbf{Qy}$ and $\mathbf{x}^* = \mathbf{Qx}$. It follows $\mathbf{x} = \mathbf{Q}^{-1}\mathbf{x}^*$ and $\mathbf{y} = \mathbf{Q}^{-1}\mathbf{y}^*$. Introducing this into Equation 18.2 yields:

$$\mathbf{Q}^T\mathbf{y}^* = \mathbf{AQ}^T\mathbf{x}^*. \tag{18.3}$$

Left-hand multiplication with $\mathbf{Q}$ results in

$$\mathbf{y}^* = \mathbf{Q}\mathbf{A}\mathbf{Q}^T\mathbf{x}^*. \tag{18.4}$$

Equating with the relation $\mathbf{y}^* = \mathbf{A}^*\mathbf{x}^*$ gives the

transformation rule for *objective tensors*:

$$\mathbf{A}^* = \mathbf{Q}\mathbf{A}\mathbf{Q}^T. \tag{18.5}$$

Instead of Equation 18.2 we can write the definition equation for the Cauchy stress tensor, $\mathbf{t} = \mathbf{T}\mathbf{n}$. It follows that $\mathbf{T}$ is an objective tensor.

18.2.2 Deformation Tensors at Change of the Reference System

Not all deformation tensors transform with rotation of the reference system according to Equation 18.5, and this means that they are not objective.

With $\mathbf{F} = \partial\mathbf{x}/\partial\mathbf{X}$ and $\mathbf{F}^* = \partial\mathbf{x}^*/\partial\mathbf{X}$ it follows from Equation 18.1

$$\mathbf{F}^*(\mathbf{X}, t) = \mathbf{Q}(t)\mathbf{F}(\mathbf{X}, t). \tag{18.6}$$

From $\mathbf{F} = \mathbf{R}\mathbf{U} = \mathbf{V}\mathbf{R}$ and $\mathbf{F}^* = \mathbf{R}^*\mathbf{U}^* = \mathbf{V}^*\mathbf{R}^*$ it follows by substituting into Equation 18.6

$$\mathbf{F}^* = \mathbf{R}^*\mathbf{U}^* = \mathbf{Q}\mathbf{F} = \mathbf{Q}\mathbf{R}\mathbf{U}, \tag{18.7}$$

hence,

$$\mathbf{R}^* = \mathbf{Q}\mathbf{R}, \quad \mathbf{U}^* = \mathbf{U}. \tag{18.8}$$

If we put the relation (Equation 18.8) into $\mathbf{V}^* = \mathbf{R}^*\mathbf{U}^*\mathbf{R}^{*T}$ we get:

$$\mathbf{V}^* = \mathbf{Q}\mathbf{R}\mathbf{U}\mathbf{R}^T\mathbf{Q}^T, \tag{18.9}$$

from which follows with $\mathbf{V} = \mathbf{R}\mathbf{U}\mathbf{R}^T$:

$$\mathbf{V}^* = \mathbf{Q}\mathbf{V}\mathbf{Q}^T. \tag{18.10}$$

That is, $\mathbf{V}$ is objective. The Cauchy–Green tensors $\mathbf{C} = \mathbf{U}^2$ and $\mathbf{B} = \mathbf{V}^2$ transform as follows:

$$\mathbf{C}^* = \mathbf{C} \tag{18.11}$$

$$\mathbf{B}^* = \mathbf{V}^{*2} = \mathbf{V}^{*T}\mathbf{V}^* = \mathbf{Q}\mathbf{V}^T\mathbf{Q}^T\mathbf{Q}\mathbf{V}\mathbf{Q}^T = \mathbf{Q}\mathbf{V}^2\mathbf{Q}^T \rightsquigarrow \mathbf{B}^* = \mathbf{Q}\mathbf{B}\mathbf{Q}^T, \tag{18.12}$$

hence, $\mathbf{B}$ is objective.

18.2.3 Transformation of v, L, D and W

How are the quantities $\mathbf{v}, \mathbf{L}, \mathbf{D}$ and $\mathbf{W}$ transformed at equivalent motions? We derive Equation 18.1 with respect to t:

$$\dot{\mathbf{x}}^*(\mathbf{X}, t) = \dot{\mathbf{q}}(t) + \mathbf{Q}(t)\,\dot{\mathbf{x}}(\mathbf{X}, t) + \dot{\mathbf{Q}}(t)\,(\mathbf{x} - \mathbf{o}).$$

With $\mathbf{v}^*(\mathbf{x}^*, t) = \dot{\mathbf{x}}^*(\mathbf{X}, t)$ and $\mathbf{v}(\mathbf{x}, t) = \dot{\mathbf{x}}(\mathbf{X}, t)$ we obtain the transformation rule for $\mathbf{v}$:

$$\mathbf{v}^*(\mathbf{x}^*, t) = \dot{\mathbf{q}}(t) + \mathbf{Q}(t)\,\mathbf{v}(\mathbf{x}, t) + \dot{\mathbf{Q}}(t)\,(\mathbf{x} - \mathbf{o}). \tag{18.13}$$

Derivation of this equation with respect to $\mathbf{x}$ gives the relationship between $\mathbf{L} = \operatorname{grad}\mathbf{v}$ and $\mathbf{L}^* = \operatorname{grad}\mathbf{v}^*$:

$$\underbrace{\frac{\partial \mathbf{v}^*}{\partial \mathbf{x}^*}}_{\mathbf{L}^*}\;\underbrace{\frac{\partial \mathbf{x}^*}{\partial \mathbf{x}}}_{\mathbf{Q}(t)} = \mathbf{Q}(t)\mathbf{L}(\mathbf{x}, t) + \dot{\mathbf{Q}}(t). \tag{18.14}$$

We multiply Equation 18.14 by $\mathbf{Q}^T$ and obtain

$$\mathbf{L}^* = \mathbf{Q}\mathbf{L}\mathbf{Q}^T + \dot{\mathbf{Q}}\mathbf{Q}^T, \tag{18.15}$$

i.e. $\mathbf{L}$ is *not* objective.

How does $\mathbf{D}$ transform? We consider that $\dot{\mathbf{Q}}\mathbf{Q}^T$ is an antimetric tensor. This results when we derive $\mathbf{Q}\mathbf{Q}^T = \mathbf{1}$ with respect to t:

$$\dot{\mathbf{Q}}\mathbf{Q}^T + \mathbf{Q}\dot{\mathbf{Q}}^T = \mathbf{0} \leadsto \dot{\mathbf{Q}}\mathbf{Q}^T = -\mathbf{Q}\dot{\mathbf{Q}}^T = -(\dot{\mathbf{Q}}\mathbf{Q}^T)^T. \tag{18.16}$$

Now we put Equation 18.15 into $\mathbf{D}^* = \frac{1}{2}(\mathbf{L}^* + \mathbf{L}^{*T})$ and obtain:

$$\mathbf{D}^* = \frac{1}{2}(\mathbf{Q}\mathbf{L}\mathbf{Q}^T + \mathbf{Q}\mathbf{L}^T\mathbf{Q}^T) = \mathbf{Q}\frac{1}{2}(\mathbf{L} + \mathbf{L}^T)\mathbf{Q}^T = \mathbf{Q}\mathbf{D}\mathbf{Q}^T, \tag{18.17}$$

hence, $\mathbf{D}$ is an objective tensor. Similarly follows:

$$\mathbf{W}^* = \mathbf{Q}\mathbf{W}\mathbf{Q}^T + \dot{\mathbf{Q}}\mathbf{Q}^T, \tag{18.18}$$

i.e. $\mathbf{W}$ is not objective.

18.2.4 Rotation of Stress

The stress can rotate in two ways:

1. The stress rotates *with* the material ('co-rotation'). The result is the same as if the observer rotates in the opposite direction. The material is *not* deformed. See Fig. 18.3b.
2. The stress rotates, but the material is not rotated, it deforms. See Fig. 18.3a.

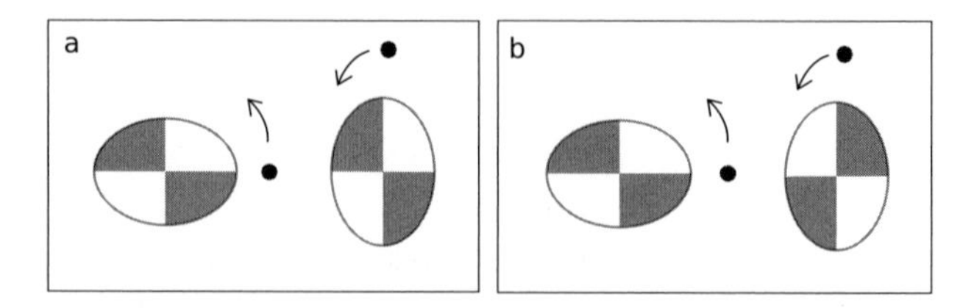

Figure 18.3 (a) The deformation of the Earth imposed by the gravity of the rotating Moon (tidal deformation). (b) Shows the Earth rotating *with* the Moon. This synchronous rotation is called tidal locking. With time, tidal deformation changes to tidal locking. Assuming a varying degree of tidal locking along the Earth's radius could explain the motion of magma in the Earth's mantle.

18.2.5 Objective Stress Rates

An observer rotating with the frame of reference registers the stress $\mathbf{T}^* = \mathbf{QTQ}^T$. If the observer is at rest and the body is rotating with $\mathbf{R}$, then the observer registers the stress $\mathbf{T}^* = \mathbf{R}^T\mathbf{TR}$. $\mathbf{T}^*$ is therefore the stress that has rotated *with* the body. The rate of $\mathbf{T}^*$ is

$$\dot{\mathbf{T}}^* = \dot{\mathbf{R}}^T\mathbf{TR} + \mathbf{R}^T\dot{\mathbf{T}}\mathbf{R} + \mathbf{R}^T\mathbf{T}\dot{\mathbf{R}}.$$

We select the current configuration as the reference configuration. This implies $\mathbf{R} = \mathbf{R}^T = \mathbf{1}$ and $\dot{\mathbf{R}} = -\dot{\mathbf{R}}^T = \mathbf{W}$. It follows:

$$\overset{\circ}{\mathbf{T}} = \dot{\mathbf{T}} - \mathbf{WT} + \mathbf{TW}.$$

$\overset{\circ}{\mathbf{T}}$ is the co-rotated or Zaremba–Jaumann stress rate and indicates the stress change due to the mechanical response of the body cleared from terms that are 'apparent', i.e. result from the change of frame.

Through the addition of objective terms, such as $\mathrm{tr}\mathbf{T} \cdot \mathbf{D}$ or $(\mathbf{DT} + \mathbf{TD})$, one can obtain further objective stress rates, e.g.

$$\hat{\mathbf{T}} = \mathbf{T} + \text{constant} \cdot \mathrm{tr}\mathbf{T} \cdot \mathbf{D}.$$

Some of the objective rates of Cauchy stress $\mathbf{T}$ proposed so far are listed in Table 18.1.

Truesdell and Noll [109] observe that: Clearly the properties of a material are independent of the choice of flux (= time rate) which, like the choice of a measure of strain, is absolutely immaterial. ... Thus we leave intentionally uncited the blossoming literature on invariant time fluxes subjected to various arbitrary requirements.

Remark Spurious oscillations at shear result from the (incorrect!) integration of $\mathbf{W}$ and $\mathbf{D}$ and are often used as an example against the Zaremba–Jaumann stress rate. Note however, that $\mathbf{D}$ and $\mathbf{W}$ refer to the current configuration and therefore integration over time may cause problems. An example is given by Belytchko et al. [7]: a cyclic deformation is divided into sections with constant $\mathbf{D}$. Integration over time results in a non-vanishing deformation! The spurious oscillation can be obtained, e.g. if we simply ignore the constitutive part of the stress rate and set $\dot{\mathbf{T}} = -\mathbf{TW} + \mathbf{WT}$. We then obtain for simple shear with $g \equiv 0$:

Table 18.1. Some co-rotated stress rates

Zaremba or Jaumann:	$\overset{\circ}{\mathbf{T}} := \dot{\mathbf{T}} - \mathbf{WT} + \mathbf{TW}$
Lie or Oldroyd:	$\mathcal{L}\mathbf{T} := \dot{\mathbf{T}} - \mathbf{LT} - \mathbf{TL}^T$
	$= \overset{\circ}{\mathbf{T}} - (\mathbf{DT} + \mathbf{TD})$
Convected stress rate:	$\mathcal{C}\mathbf{T} := \dot{\mathbf{T}} + \mathbf{L}^T\mathbf{T} + \mathbf{TL}$
	$= \overset{\circ}{\mathbf{T}} + (\mathbf{DT} + \mathbf{TD})$

$$\dot{\sigma}_{11} = \sigma_{12}\dot{f} \quad (18.19)$$
$$\dot{\sigma}_{22} = -\sigma_{12}\dot{f} \quad (18.20)$$
$$\dot{\sigma}_{12} = (\sigma_{22} - \sigma_{11})\dot{f}/2. \quad (18.21)$$

We set $\dot{f}$ = constant and derive with respect to t obtaining thus the differential equation $\ddot{\sigma}_{12} = -\dot{f}^2\sigma_{12}$ with the periodic solution $\sigma_{12} = A\cos\dot{f}t + B\sin\dot{f}t$.

18.3 Isotropic Materials

Isotropy means invariance with respect to rotations. Isotropy of space means that the physical processes are the same in all directions. Similarly, homogeneity of space means that the physical processes are the same if we consider them in a translated position. Homogeneity and isotropy of space are important properties, from which the conservation of momentum and angular momentum can be derived [64]. Since the notion of space is inherently related with the notion of a frame (relative to which we can realise motions), the here stated invariance is related to the principle of frame indifference, which is important for the formulation of constitutive equations. In this section we consider isotropy with respect to material properties.

Isotropic materials exhibit the same properties in all directions. In contrast, there are materials which exhibit a mechanically preferred orientation. This is the case, for example, with metamorphic rock and with overconsolidated clay, which is stiffer in the direction of the maximum consolidation stress component.

There are many different anisotropic solids and they are categorised with the rotations that cannot be mechanically detected. Regarding the aforementioned overconsolidated clay, rotations about the direction of maximum applied consolidation stress cannot be mechanically detected, and such materials are called cross anisotropic.

All motions $\mathbf{B}$ which remain mechanically undetectable in subsequent deformations, such that

$$\mathcal{G}_{s=0}^{\infty}(\mathbf{F}^{(t)}(s)) = \mathcal{G}_{s=0}^{\infty}(\mathbf{F}^{(t)}(s)\mathbf{B}),$$

form the *isotropy group* or *symmetry group* of a material. This property applies with respect to a particular reference configuration. Herein, the symbol $\mathcal{G}_{s=0}^{\infty}(\mathbf{F}^{(t)}(s))$ simply denotes a particular (not further specified) history of deformation that extends from $t = -\infty$ to the present ($t = 0$).

To add another example, the isotropy group of simple fluids is the group of all volume-preserving deformations. This implies that there is no way to detect whether water has been sheared in the past.

If an orthogonal transformation $\mathbf{Q}$ belongs to the isotropy group of a material, it follows from the principle of frame indifference,

$$\mathbf{Q}_0\mathcal{G}_{s=0}^{\infty}(\mathbf{F}^{(t)}(s))\mathbf{Q}_0^T = \mathcal{G}_{s=0}^{\infty}(\mathbf{Q}(s)\mathbf{F}^{(t)}(s)), \qquad \text{with} \quad \mathbf{Q}_0 := \mathbf{Q}(t = 0).$$

A material is *isotropic* if at least one local reference configuration (so-called undistorted state) exists, with respect to which the isotropy group is the full orthogonal group. For isotropic materials, previous rotations cannot be detected by mechanical experiments. The isotropy group of anisotropic solids is a subgroup of the full orthogonal group.

18.3.1 Cross Anisotropy

Cross anisotropy is encountered e.g. in highly overconsolidated clay (parallel arrangement of platelets), in stratified sedimentary rocks and in shaled metamorphic rocks. The mechanical behaviour of cross-anisotropic materials does not change (i.e. remains mechanically undetectable) when rotating around a vector $\mathbf{n}$, which is perpendicular to the platelets (in the case of overconsolidated clay) or to the slate planes (in case of metamorphic rocks). For cross anisotropy, the linear elastic law has six independent material constants. As in the isotropic case, different formulations are possible. Statements in the literature usually refer to special alignments of $\mathbf{n}$ with respect to the underlying coordinate system. For a linear elastic, cross-anisotropic material with general orientation of $\mathbf{n}$, the constitutive relation reads [48]:

$$\mathbf{T} = (a_1 \mathrm{tr}\mathbf{E} + a_2 \mathbf{n} \cdot \mathbf{En})\mathbf{1} + a_3 \mathbf{E} + (a_4 \mathrm{tr}\mathbf{E} + a_5 \mathbf{n} \cdot \mathbf{En})\ \mathbf{n} \otimes \mathbf{n} + a_6 (\mathbf{n} \otimes \mathbf{En} + \mathbf{En} \otimes \mathbf{n}). \tag{18.22}$$

Remark Equation 18.22 is written in symbolic notation. It can be easily transferred into the index notation, which is more appropriate for programming. Noting that the dyadic product $\mathbf{n} \otimes \mathbf{n}$ can be also written as $n_i n_j$, we obtain

$$T_{ij} = (a_1 E_{kk} + a_2 n_k E_{kl} n_l)\delta ij + a_3 E_{ij} + (a_4 E_{kk} + a_5 n_k E_{kl} n_l) n_i n_j + a_6 (n_i E_{jk} n_k + E_{ik} n_k n_j). \tag{18.23}$$

18.4 Scaling

Changes of scale may affect the observed behaviour of a material. The following scales can be considered:

- Change of geometric scale: It matters for so-called non-simple materials ('size effect'), see Section 18.4.1.
- Change of time scale: It matters for so-called rate-dependent materials.
- Change of stress scale: It is immaterial if the constitutive relation $\dot{\mathbf{T}} = \mathbf{h}(\mathbf{T}, \mathbf{D})$ is homogeneous in degree zero with respect to $\mathbf{T}$. This means that the mechanical behaviour is invariant with respect to changes of the stress scale.

18.4.1 Invariance with Respect to Geometric Scale, Simple Materials

Most constitutive equations are tacitly based on the assumption of the so-called simple material, according to which only the first deformation gradient and its history is decisive for the current stress at a point on a continuum [109]. This implies that element tests are sufficient for calibration of a constitutive equation and that the material has no internal length. Therefore, two triaxial experiments on geometrically similar samples of different sizes (for example with the diameters 10 cm and 100 cm) of the same material and with the same boundary conditions (i.e. with the same lateral pressure) result in identical stress–strain curves.

Basically, *the entire procedure in soil mechanics is based on the assumption that soil is a simple material*. The extracting of soil samples, their laboratory examination and the subsequent computations are based thereupon. Virtually, the invariance with respect to geometric scale is not strictly applied and a certain scale effect is always observed, but it can be disregarded as long as the considered soil samples are large compared to their grains.

In rock mechanics, however, the size effect is considerable, mainly due to the existence of discontinuity surfaces (e.g. joints), and this makes the application of continuum mechanics questionable for jointed rock. Thus, the calculation methods of rock mechanics are – strictly speaking – only applicable to unjointed rock, which is rather rare. Admittedly, there is a makeshift that finds wide use: continuum mechanics-based calculations are carried out using a relation by Hoek and Brown that suggests empirical values for the shear strength parameters φ and c. Despite its wide application, this method is only a rough workaround.

18.4.2 Invariance with Respect to Stress Scale

In model tests that are scaled down by the factor N, the gravity-induced stresses are scaled by the same factor. If the stiffness of the material (which is assumed to be the same in the model and the prototype) is also scaled by N, then the dimensionless strains resulting from whatever loading will be the same in both the prototype and the model, the considered soil is called 'mechanically self-similar'. This is the case for soils described by a constitutive equation $\dot{\mathbf{T}} = \mathbf{h}(\mathbf{T}, \mathbf{D}, \ldots)$ that is homogeneous of in the first degree in $\mathbf{T}$, i.e. $\mathbf{h}(\lambda\mathbf{T}, \mathbf{D}, \ldots) = \lambda\mathbf{h}(\mathbf{T}, \mathbf{D}, \ldots)$ for $\lambda > 0$. This implies that stiffness is proportional to the actual stress level. Real soils however are not exactly self-similar, their stiffness increases underlinearly with stress level.

18.4.3 Invariance with Respect to Time Scale, Rate Independence

Invariance with respect to the time scale is given when the constitutive equation is homogeneous in the first degree in $\mathbf{D}$, i.e. $\mathbf{h}(\mathbf{T}, \lambda\mathbf{D}, \ldots) = \lambda\mathbf{h}(\mathbf{T}, \mathbf{D}, \ldots)$ for $\lambda > 0$. In this case, no material constant with the dimension of time appears in the constitutive equation, and the stress resulting from a given deformation is independent of the rate with which this deformation occurred. Such materials are called rate independent. Elastic, elastoplastic, hypoplastic and barodetic constitutive equations without appropriate extensions are rate independent, but real materials exhibit some rate dependence.

The viscosity known from fluids means that the resistance to shear increases with the shear rate. For solids it is often assumed that the speed of deformation does not play any role (rate independence). This means that we get the same stress–strain curve regardless of how fast we deform a sample. However, soil (and rock) is rate independent only to a first approximation, and there are many cases where rate dependence plays a role. As a consequence, the stress–strain curve, e.g. from a strain-controlled triaxial test, will depend on whether the test is performed at the constant deformation rate $\dot{\varepsilon}_0$ or at $\dot{\varepsilon}_1 \neq \dot{\varepsilon}_0$. However, since the difference between the

Table 18.2. Typical values of I_v

Sand	0.01
Silt	0.02
Clay, medium plasticity	0.03
Clay, high plasticity	0.04
Organic mud	0.05

two curves is quite small and might be lost in the general scatter of the experimental results, to detect rate dependence it is advisable to perform an experiment with an abrupt change of the deformation rate from $\dot{\varepsilon}_0$ to $\dot{\varepsilon}_1$. The observed stress jump (Fig. 2.25) is related to the change of strain rate according to:

$$\Delta\tau = \tau - \tau_0 = I_v \tau_0 \Delta(\ln\dot{\varepsilon}) = I_v \tau_0 \ln\left(\frac{\dot{\varepsilon}_1}{\dot{\varepsilon}_0}\right), \tag{18.24}$$

where I_v is the so-called viscosity index [66]. The value of I_v, and thus the importance of the rate sensitivity, is greater for clay than for sand, as shown in Table 18.2 [60].

Remarks:

1. In Equation 18.24, τ stands for the stress deviator $\sigma_1 - \sigma_2$ or for a shear stress. However, it is more realistic to formulate it for the ratio τ/σ' (or for $(\sigma_1'-\sigma_2')/(\sigma_1'+\sigma_2')$), where σ' is the effective normal stress [32].

$$\Delta\left(\frac{\tau}{\sigma'}\right) = \frac{\tau_0}{\sigma_0'} - \frac{\tau}{\sigma'} = I_v \cdot \frac{\tau_0}{\sigma_0'} \cdot \Delta\left(\ln\dot{\varepsilon}\right) = I_v \cdot \frac{\tau_0}{\sigma_0'} \cdot \ln\frac{\dot{\varepsilon}}{\dot{\varepsilon}_0}. \tag{18.25}$$

   This equation makes it possible to describe the accelerated creep of slopes when the groundwater rises. In this case, the shear stress is reduced, but the ratio of the shear stress to the normal stress is increased.
2. The logarithmic viscosity law (Equation 18.24) can also be formulated as a power law. Since for small x applies $1 + \ln x \approx x$, this equation can also be formulated as:

$$\frac{\tau}{\tau_0} = 1 + I_v \ln\frac{\dot{\varepsilon}}{\dot{\varepsilon}_0} = 1 + \ln\left(\frac{\dot{\varepsilon}}{\dot{\varepsilon}_0}\right)^{I_v} \approx \left(\frac{\dot{\varepsilon}}{\dot{\varepsilon}_0}\right)^{I_v}. \tag{18.26}$$

3. The rate dependence according to Equation 18.25 is closely related to creep and relaxation. Observation shows that relaxation proceeds logarithmically with time, and that the creep rate often decreases with time. Accelerated creep leads eventually to rupture [25, 61].
4. The rate sensitivity of sand is more complicated than according to Equation 18.25, see Fig. 2.26.
5. Equation 18.25 can also be applied to other boundary value problems with rate-dependent materials. For example, with Q being the load of a pile and v the speed of penetration, we have a dependence of Q on v of the type $Q = Q_0 \ln(v/v_0)$.

Remark The power law $\tau = \mathrm{const}\dot{\varepsilon}^{\alpha}$ (see Equation 18.26) is known as the law of Norton or Glen. In glaciology it is also known as the law of Weertmann. A relationship similar to Equation 18.25 was also found for the friction in rock joints by Dieterich and Ruina [89].

From microscopic considerations, Prandtl concluded a logarithmic dependence of the sliding friction force on the velocity and mentioned experiments with sliding friction between brass and glass as well as leather and iron [88]. A similar theory became known later as 'rate process theory' [28]. According to this theory, creep occurs due to the thermally activated overcoming of energy barriers. For such processes the equation of Arrhenius is valid: $\dot{\varepsilon} \sim \exp(-Q/RT)$, where Q and R are constants (R is the gas constant, $R = 8.314472$ J/(mol·K), and T is the absolute temperature). Hence, the creep rate increases with temperature.

18.5 Mechanical Similarity

Physical modelling is an alternative to numerical simulation. Reducing the geometric scale of a geotechnical construction (e.g. the foundation of a bridge) enables us to investigate its behaviour (i.e. its deformation) in the laboratory with reduced costs and risk. Of course, we have to ensure that both the model and the prototype are mechanically (not only geometrically) *similar*.

It is often useful to combine the relevant variables of a system into dimensionless variables Π_i. This has the advantage of lowering the number of variables. For example, the number of variables in the case of the elongation of a string of length l can be reduced according to Hooke's law:

$$f(F, A, E, \Delta l, l) = 0 \quad \rightarrow \quad f\left(\frac{\sigma}{E}, \varepsilon\right) = 0. \tag{18.27}$$

Moreover, it allows the following definition of mechanical similarity:

Two problems are similar when all dimensionless variables have the same values in both the model and the prototype.

A mathematical relationship between dimensionless variables describes not only one problem, but a class of similar problems. For example, the deformation of a structure ('prototype') can be analysed by reconstructing it on a geometrically reduced scale ('model') and then investigating it in the laboratory. This procedure is also called 'physical simulation'. With its help, numerical simulations can be circumvented, checked or calibrated.

18.5.1 Π-Theorem

From a set of original variables, we can always get a set of dimensionless variables (so-called Π-theorem, Vaschy 1892, Buckingham 1914). This is a result of the requirement that physical laws be invariant with respect to a change of the units. For example, the formula force = mass × acceleration is valid, no matter whether you measure acceleration in m/s^2 or ft/s^2.

Mechanical similarity on the basis of dimensionless variables can be explained by the following examples:

Mathematical pendulum: We look for a relationship between the frequency ω, the pendulum length l, the pendulum mass m and the acceleration due to gravity g of the form $F(\omega, l, m, g) = 0$. We see that the mass m cannot be non-dimensionalised by combining it with any of the other variables, so m is not a variable of the problem. The only dimensionless variable that can be formed with the remaining variables is $\Pi := \frac{\omega^2 l}{g}$. From this it follows that the searched relationship has the form $F_1(\Pi) = 0$, from which in turn follows $\Pi = \text{const}$ or $\omega = \text{const} \cdot \sqrt{g/l}$. The value of the constant can be experimentally determined.

Permeability: An example of a dimensionally incorrect formula is a formula of Hazen for the permeability (see Section 19.15): $k = 0.01 \cdot D_{10}^2$. Herein, D_{10} is the grain diameter such that grains belonging to 10% of the weight of a sample have a smaller diameter. Such formulas are only useful if the units are specified. For example, in the above formula, D_{10} is expressed in mm and k is expressed in m/s. However, it is easy to use dimensional analysis to create a dimensionally correct formula. Note that the permeability $k(:= v/i)$ depends not only on the grain skeleton, but also on the acceleration of gravity g, since the pressure is measured in metre water column (on the Moon k has a different value than on the Earth's surface!). Directly related to the material is the quantity $\bar{k} := v/\nabla p = k/\gamma_w$. Now $\bar{k}$ depends on an effective grain diameter D_w (which represents a suitable diameter of a pore channel) and the viscosity μ of the fluid. From $F(k/\gamma_w, \mu, D_w) = 0$ follows $F_1(\Pi) = 0$ with $\Pi = \dfrac{k\mu}{\gamma_w D_w^2}$. It follows: $\Pi = \text{const}$ or $k = \text{const} \cdot D_w^2 \gamma_w / \mu$.

Hourglass: From experience we know that the run-out speed v from an hourglass does not depend on the filling height h. The searched relationship has therefore the form $F(d, v, g) = 0$, where d is the diameter of the outlet and g is the acceleration of gravity. With $\Pi = v^2/(dg)$ the relationship takes the form $F_1(\Pi) = 0$. From this follows $\Pi = \text{const}$ or $v = \text{const} \times \sqrt{d \cdot g}$ and $Q = (\pi d^2/4) \cdot v = c_1 \cdot d^{5/2}$. It should be added that related experiments exhibit a large scatter.

Consolidation: In one-dimensional consolidation (see Section 19.6), the significant variables are related by an equation $F(s, d, k/\gamma_w, \delta p, E_s, t) = 0$. From this, we can derive the following dimensionless variables:

$$\Pi_1 = s/d; \quad \Pi_2 = \frac{k \cdot t \cdot E_s}{\gamma_w \cdot d^2}; \quad \Pi_3 = \frac{\delta p}{E_s}.$$

Here, δp is the load, s the settlement and d the thickness of the considered soil layer. If the same material is used in both the model and the prototype with the same load, it follows from $\Pi_{2model} = \Pi_{2prototype}$:

$$\left(\frac{t}{d^2}\right)_{model} = \left(\frac{t}{d^2}\right)_{prototype}.$$

This implies that the time needed for a particular degree of consolidation increases with the square of the thickness d of the considered soil layer.

18.5.2 Centrifuges

Centrifuges enable model tests to be carried out with increased gravity (Fig. 18.4). Depending on the radius r and angular velocity ω, the centrifugal acceleration $\omega^2 r$ can exceed the acceleration of gravity by a considerable factor.

A model experiment in the centrifuge is useful if we want, for example, to check the stability of an embankment in cohesive soil. From the theoretical analysis (or

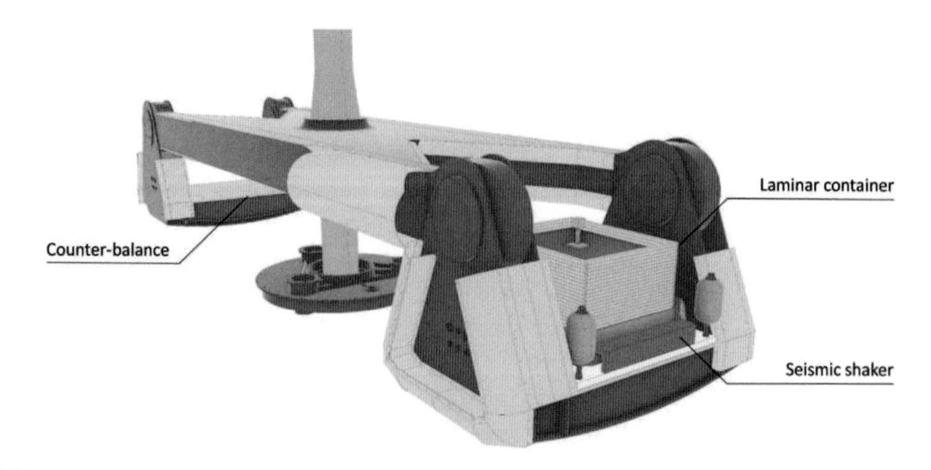

Figure 18.4 Geotechnical centrifuge of the ETH Zürich, courtesy of Professor I. Anastasopoulos.

from the Π-theorem) we know that the dimensionless variable $\frac{\gamma h}{c}$ plays a role and therefore it should match in both prototype and model. If we now reduce the geometrical scale by the factor $1/N$, i.e. $h_M = h_P/N$ (h_M = height of the embankment in the model, h_P = height of the embankment in the prototype), then either c must be decreased by the factor $1/N$ or γ increased by the factor N. To downsize cohesion is difficult, as it means that we have to use a material, which may have completely different properties regarding dilatancy, stiffness, etc. Therefore, it is advantageous, to use the same soil and increase the gravity g in the centrifuge.

Errors occur in centrifuge tests due to Coriolis forces (if radial velocities occur during the test) and due to the fact that the mass force is not homogeneous, but increases with the radius r. Also, the often necessary strong miniaturisation of construction processes poses problems.

18.5.3 Mathematics behind the Π-Theorem

A physical problem may be described by n variables $x_1, x_2, \ldots x_n$. Each variable x_i is represented as product of a dimensionless number ξ_i and some powers of basic units E_i:

$$x_i = \xi_i E_1^{e_{i1}} E_2^{e_{i2}} \ldots E_k^{e_{ik}},$$

where k is the number of basic units that appear in this problem. Instead of the n dimensional variables $x_1, x_2, \ldots x_n$ we now use the dimensionless variables $\Pi_1, \Pi_2, \ldots$. Each dimensionless variable is the product of powers of the variables x_i:

$$\begin{aligned} x_1^{p_1} \cdot x_2^{p_2} \cdot \cdots \cdot x_n^{p_n} = \; & \xi_1^{p_1} \cdot E_1^{e_{11}p_1} \cdot E_2^{e_{12}p_1} \cdot \cdots \cdot E_k^{e_{1k}p_1} \times \\ & \xi_2^{p_2} \cdot E_1^{e_{21}p_2} \cdot E_2^{e_{22}p_2} \cdot \cdots \cdot E_k^{e_{2k}p_2} \times \\ & \ldots \\ & \xi_n^{p_n} \cdot E_1^{e_{n1}p_n} \cdot E_2^{e_{n2}p_n} \cdot \cdots \cdot E_k^{e_{nk}p_n}. \end{aligned}$$

Since the new variables are dimensionless, it must follow that:

$$\begin{aligned} e_{11}p_1 + e_{21}p_2 + \cdots + e_{n1}p_n &= 0, \\ e_{12}p_1 + e_{22}p_2 + \cdots + e_{n2}p_n &= 0, \\ &\cdots \\ e_{1k}p_1 + e_{2k}p_2 + \cdots + e_{nk}p_n &= 0. \end{aligned}$$

So, there is a homogeneous system of k equations with n unknowns $p_1, p_2, \ldots, p_n$. Such a system has m linear independent solution vectors:

$$\begin{pmatrix} p_1 \\ p_2 \\ \vdots \\ p_n \end{pmatrix}_1, \begin{pmatrix} p_1 \\ p_2 \\ \vdots \\ p_n \end{pmatrix}_2, \ldots \begin{pmatrix} p_1 \\ p_2 \\ \vdots \\ p_n \end{pmatrix}_m,$$

and each of these m vectors determines a dimensionless variable Π_i. The number m follows from the rank r of the system of equations. r is given by the fact that no non-vanishing subdeterminant of order $r+1$ exists. Hence, $m = n - r$. Practically, the dimensionless variables Π_i are found by trial and error, and there are generally many different sets of dimensionless variables. The selected variables should correlate with each other as well as possible. The choice of the most favourable set of variables depends on skill.

Mechanical similarity exists if all dimensionless variables Π_i have the same value in both the prototype and the model. In most cases, this requirement cannot be strictly met.

19 Interaction with Water

19.1 Water in Soil

The interaction of the soil grains with the water contained in their pores causes important and often peculiar effects. Most of the damage in geotechnical engineering is due to the action of water.

19.2 Multiphase Materials

Materials consisting of more than one components are called composite or multiphase materials or mixtures. As each of their constituents occupies many small parts of the space, it makes no sense to consider them separately. Instead, it is assumed that each constituent is smoothly distributed over the entire space of the considered mixture. Thus, every point is occupied by several phases. For each constituent the conservation laws of mechanics are held. In addition, we have to take into account the interaction of the individual components. A pronounced multiphase material is water-saturated soil, which consists of grains and pore water.

An important notion for multiphase materials is the so-called volume fraction α_i of the constituent i: $\alpha_i := V_i/V$, with V_i = volume of fraction i and V = total volume. Clearly, the sum over all phases equals 1, $\sum \alpha_i = 1$. The volumes V_i and V should be considered over a region that is not too small (so that it contains representative parts of each phase) but also not too large (so that the spatial variability of each phase can still be considered). For water-saturated soil we have

$$\alpha_w = n, \quad \alpha_s = 1 - n, \tag{19.1}$$

for water and solids, respectively. n is the porosity.

Analogously to the volume fractions we can also consider area fractions, for which the following theorem holds:

Theorem of Delesse: The volume fraction of pores, α_w or n, equals the area fraction of pores $n_A := A_w/A$:

$$n_A = n. \tag{19.2}$$

Proof: The volume V of a soil body results from the integration of its cross-sectional area A over a perpendicular coordinate x with $V = \int_{x_1}^{x_2} A(x)\mathrm{d}x$. In the same way the pore volume is obtained as $V_p = \int_{x_1}^{x_2} n_A A(x)\,\mathrm{d}x$. If the body is statistically homogeneous, then n_A is independent of x and can be factored out: $V_p = n_A \int_{x_1}^{x_2} A(x)\,\mathrm{d}x = n_A V$, from which the equality $n_A = n$ follows. In the same way it can be shown that the line porosity is equal to the area porosity. For the velocity of pore water we have: $v^w = v_w/n$.

19.2.1 Partial versus True Quantities

Field quantities, such as density ρ, are assigned to each phase and are displayed with the corresponding index. We should distinguish between partial and real (or true) quantities. For example, ρ_s is the partial density of the soil grain skeleton, and $\rho^s (\approx 2.7\ \mathrm{g/cm^3}$, typically) is the real or true density of a grain. The partial quantities can be obtained from the real quantities by multiplication with the corresponding volume fraction, e.g.

$$\rho_s = \alpha_s \rho^s. \tag{19.3}$$

The total density of the mixture equals the sum of the partial densities of the individual phases:

$$\rho = \rho_s + \rho_w. \tag{19.4}$$

In the same way we distinguish between the partial velocity of pore water v_w (which is also called 'filter velocity' and often denoted as v) and the real velocity v^w. The stretching $\mathbf{D}$ of the grain skeleton is obtained with its true velocity, $D_{ij} = (v^s_{i,j} + v^s_{j,i})/2$. For real and partial field quantities, the indices w and s are used as superscript or subscript, respectively.

19.2.2 Internal Constraints to Deformation

Soil can be mixed with materials that impose constraints (also called 'auxiliary conditions', see [65]) to its deformation. Such materials can be water or inextensible fibres. The latter are often assumed as smeared to render a homogeneous two-phase material. Internal constraints limit the motion by an imposed kinematic condition

$$f(\mathbf{D}) = 0. \tag{19.5}$$

To maintain the constraint, a stress $\mathbf{N}$ is required which does not do any work, i.e.

$$\mathrm{tr}(\mathbf{ND}) = 0. \tag{19.6}$$

The constraint must be *rate independent*, i.e. the function $f(\mathbf{D})$ must be homogeneous in the first order in $\mathbf{D}$. According to Euler's theorem it must therefore apply:

$$f = \frac{\partial f}{\partial D_1} D_1 + \frac{\partial f}{\partial D_2} D_2 + \cdots = \mathrm{tr}\left(\frac{\partial f}{\partial \mathbf{D}} \mathbf{D}\right) = 0. \tag{19.7}$$

In terms of geometry, Equations 19.6 and 19.7 imply that $\mathbf{N}$ is perpendicular to all $\mathbf{D}$ that are perpendicular to $\frac{\partial f}{\partial \mathbf{D}}$. Hence, $\mathbf{N}$ and $\frac{\partial f}{\partial \mathbf{D}}$ are parallel:

$$\mathbf{N} = \lambda \frac{\partial f}{\partial \mathbf{D}}. \tag{19.8}$$

The scalar multiplier λ is constitutively undetermined and must be determined from a static boundary condition.

Examples of Constraints

Incompressibility: From the condition for *isochoric* deformation, $\mathrm{tr}\mathbf{D} = D_{ii} = 0$, follows

$$\mathbf{N} = \lambda \frac{\partial D_{ii}}{\partial D_{kl}} = \lambda \delta_{ik}\delta_{il} = \lambda \delta_{kl} = \lambda \mathbf{1},$$

i.e. $\mathbf{N}$ is a hydrostatic stress. Water has a very low compressibility. Thus, a water-saturated soil can be considered as incompressible, if drainage is inhibited ('undrained soil'), and because soil grains can also be considered as incompressible. In this case, the multiplier λ is the pore pressure p. If we consider an undrained triaxial test, the static boundary condition requires that the total lateral stress remains constant and equals the sum of effective lateral stress plus pore pressure: $\sigma_2 = \sigma_2' + p = \text{const}$, see also Section 19.3. In general, the stress in water-saturated soil is the sum of a constitutively (i.e. due to deformation) caused part plus a hydrostatic part that counts for the incompressibility.

As usual in soil mechanics, the symbol p denotes here the pore pressure. Elsewhere in this book it denotes the mean stress, $p = (\sigma_1 + \sigma_2 + \sigma_3)/3$. This is admittedly confusing, but it is hoped that the reader will infer the proper meaning from the context.

Inextensibility: Inextensibility in the direction of the unit vector $\mathbf{e}$, see Equation 6.16, induced for instance by rigid steel inlets (soil nailing), implies:

$$f(\mathbf{D}) = \mathbf{e} \cdot \mathbf{D}\mathbf{e} = D_{ij}e_ie_j = 0.$$

It then follows:

$$\mathbf{N} = \lambda \frac{\partial (D_{ij}e_ie_j)}{\partial D_{kl}} = \lambda\ \delta_{ik}\ \delta_{il}\ e_i\ e_j = \lambda\ e_k\ e_l = \lambda\ \mathbf{e} \otimes \mathbf{e}.$$

Nailing, i.e. driving of steel rods into the soil, renders the resulting composite material more or less inextensible in the direction of the nails.

Rigidity: The entire stress tensor $\mathbf{T}$ is constitutively indefinite.

19.3 Effective Stress in Water-saturated Soil

We consider a cross section with unit normal vector $\mathbf{n}$ and decompose the total stress vector $\mathbf{t}$ acting thereupon into the partial stress vectors of grain skeleton and pore water:

$$\mathbf{t} = \mathbf{t}_s + \mathbf{t}_w. \tag{19.9}$$

The partial stress vector in the pore water is $\mathbf{t}_w = np\mathbf{n}$, hence the partial pressure in the pore water is $p_w = np$, where $p\ (= p^w)$ is the real pressure in the pore water, the so-called pore-water pressure. The partial stress in the grain skeleton is then $\sigma_{ij,s} = \sigma_{ij} - np\delta_{ij}$. Note that the hydrostatic pressure p also acts upon the individual grains. Thus, the grain skeleton carries the part $(1 - n)p$ of the pore pressure p. As the grains of common soils are incompressible, this part has no influence on the behaviour of the grain skeleton. The latter is therefore only affected by the difference, the so-called effective stress $\sigma'_{ij} := \sigma_{ij,s} - (1-n)p\delta_{ij} = \sigma_{ij} - p\delta_{ij}$ (principle of effective stress). Hence, constitutive equations link the deformation of the grain skeleton with the *effective* stress. In previous sections the effective stress was always meant, even if the dash was omitted for simplicity, because for dry or drained soil (i.e. for $p = 0$) $\sigma' = \sigma$ applies.

It is often considered as a necessary condition for the validity of the principle of effective stress, that the grain contacts have vanishing small surfaces. However, we have seen that its validity depends only on the incompressibility of the grains. It should be added that for not negligible grain contact areas, a change of the pore pressure affects the forces transmitted from grain to grain. The associated macroscopic stress change does not cause any deformation, but it can slightly increase the shear strength of the grain skeleton [95].

In the case of (i) undrained conditions, (ii) incompressible grains and (iii) compressible pore fluid, the principle of effective stresses still applies, but now we need to know how a total stress $\Delta\sigma$ applied upon the boundary is partitioned between the pore fluid and the grain skeleton. The pore water is compressible if it contains air bubbles, as is often the case. With the compressibility κ of the pore fluid and the pore volume V_p, we obtain the change of pore volume as $\Delta V_p = -\kappa V_p \Delta p$. Hence, the change of the void ratio is

$$\Delta e = \frac{\Delta V_p}{V_s} = \frac{\kappa V_p \Delta p}{V_s} = -\kappa e \Delta p. \tag{19.10}$$

On the other hand, the compressibility of the grain skeleton implies (Equation 2.1):

$$\Delta e = -C_c \frac{\Delta\sigma'}{\sigma'}. \tag{19.11}$$

Equating Equations 19.10 and 19.11 yields $\Delta\sigma' = \frac{\sigma'\kappa e}{C_c}\Delta p$. From $\Delta\sigma = \Delta\sigma' + \Delta p$ it then follows $\Delta\sigma = \left(1 + \frac{\sigma'\kappa e}{C_c}\right)\Delta p$ or

$$\Delta p = B\Delta\sigma \quad \text{with} \quad B = \frac{1}{1 + \frac{\sigma'\kappa e}{C_c}}. \tag{19.12}$$

For incompressible pore fluid (i.e. $\kappa = 0$) we have $B = 1$.

19.4 Darcy's Law

If the pore water and the grains are at rest, then the distribution of the pore pressure is hydrostatic: $p/\gamma^w = h - z$, where $\gamma^w = \rho^w g$ and z is the vertical coordinate denoting the distance from the water table ($z = h$), which is horizontal in this case. It follows: $\nabla h = \mathbf{0}$, where $h := p/\gamma^w + z$ is the so-called energy head, a quantity useful in hydraulic calculations. According to this equation, the energy head h consists of the pressure head p/γ_w and the geometric head z. The so-called velocity head $(v^w)^2/(2g)$, considered in hydraulics, is negligibly small in groundwater flow. Obviously, $\nabla h \neq \mathbf{0}$ holds true whenever the pore water moves relative to the grain skeleton, i.e. when $\mathbf{v}^w - \mathbf{v}^s \neq \mathbf{0}$, hence a relation between ∇h and $\mathbf{v}^w - \mathbf{v}^s$ must exist. Herein, the quantities γ^w, d (a representative pore diameter) and μ (viscosity of pore fluid) must also play a role. So, this relation has the general form

$$F(\gamma^w, \nabla h, \mathbf{v}^w - \mathbf{v}^s, d, \mu) = 0. \tag{19.13}$$

If we consider the values of the vectors $\mathbf{v}^w - \mathbf{v}^s$ and ∇h, we obtain the dimensionless variable $\Pi := \frac{(v^w - v^s)\mu}{\gamma^w \nabla h\, d^2}$, so that Equation 19.13 takes the form $F(\Pi) = 0$, hence $\Pi = \text{const}$, i.e.

$$\mathbf{v}^w - \mathbf{v}^s = \text{const} \cdot \frac{\gamma^w d^2}{\mu} \nabla h. \tag{19.14}$$

For the special case $\mathbf{v}^s = \mathbf{0}$ and with $\mathbf{v} = \mathbf{v}_w = n\mathbf{v}^w$, Darcy's equation is obtained:

$$\mathbf{v} = -k\nabla h \tag{19.15}$$

which in the case of moving grains takes the more general form (Darcy–Gersevanov's equation):

$$\mathbf{v} - n\mathbf{v}^s = -k\nabla h, \tag{19.16}$$

with the so-called coefficient of permeability

$$k := \text{const} \cdot \frac{n\rho^w g d^2}{\mu}. \tag{19.17}$$

19.5 Balance Equations

For multiphase materials, the balance equations are formulated for the individual constituents separately. Here, only the quasi-static case is considered, therefore the impulse balance reduces to the equilibrium equation, which has different versions referring either to true or to partial quantities. In addition, the equilibrium equation can also be formulated in terms of effective stress. For non-negligible accelerations, the interaction between the constituents is complex and omitted here.

19.5.1 Balance Equations for Partial Quantities

The balance equations (Equations 19.18–19.21) for the individual phases have the same structure as for single-phase materials with the only difference being that in the equilibrium equations the interaction force $\mathbf{R}$ now appears, which describes how one phase acts upon the other.

Mass balance, pore water: $$\partial_t \rho_w + \partial_i(\rho_w v_i^w) = 0, \tag{19.18}$$

Mass balance, grain skeleton: $$\partial_t \rho_s + \partial_i(\rho_s v_i^s) = 0, \tag{19.19}$$

Equilibrium, pore water: $$-\partial_i p_w + \rho_w b_i^w - (R_{ws})_i = 0, \tag{19.20}$$

Equilibrium, grain skeleton: $$-\partial_j \sigma_{ij,s} + \rho_s b_i^s + (R_{ws})_i = 0. \tag{19.21}$$

Here, the symbol ∂ with the subscripts t and i indicates the partial derivatives with respect to t and x_i. b_i^w and b_i^s are the mass forces acting upon the pore water and grain skeleton, respectively. They result from gravity, i.e. $b_i^t = b_i^s = g_i$, where g_i is the acceleration due to gravity, or from other accelerations. The volume force $\mathbf{R}_{ws}$ represents the action of pore water upon the grain skeleton. Clearly, $\mathbf{R}_{ws} = -\mathbf{R}_{sw}$.

19.5.2 Balance Equations for True Quantities

With ρ^w = const and ρ^s = const we obtain from Equations 19.18–19.21:

Mass balance, pore water:
$$\partial_t n + v_i^w \partial_i n + n\partial_i v_i^w = 0, \tag{19.22}$$

Mass balance, grain skeleton:
$$-\partial_t n - v_i^s \partial_i n + (1-n)\partial_i v_i^s = 0, \tag{19.23}$$

Equilibrium, pore water:
$$-p^w \partial_i n - n\partial_i p^w + n\rho^w b_i^w - R_i^{ws} = 0, \tag{19.24}$$

Equilibrium, grain skeleton:
$$-(1-n)\partial_j \sigma_{ij}^s + \sigma_{ij}^s \partial_j n + (1-n)\rho^s b_i^s + R_i^{ws} = 0. \tag{19.25}$$

19.5.3 Interaction force

The interaction force $\mathbf{R}$ is a volume force, i.e. a force per unit volume. Comparing Equations 19.16 and 19.24 we obtain with $\partial_i p^w - \rho^w g_i = \partial_i(p^w + \gamma^w z) = \gamma^w \partial_i h$:

$$\mathbf{R}^{ws} = \frac{n\gamma^w}{k}(\mathbf{v} - n\mathbf{v}^s) - p^w \nabla n = -n\gamma^w \nabla h - p^w \nabla n. \tag{19.26}$$

From Equation 19.20 we obtain:

$$\mathbf{R}_{ws} = -n\gamma^w \nabla h - p^w \nabla n = \mathbf{R}^{ws}. \tag{19.27}$$

In the case of acceleration, the interaction between grains and pore fluid is more complex.

19.5.4 Equilibrium in Terms of Effective Stresses

With the aforementioned equation $\sigma'_{ij} := \sigma_{ij,s} - (1-n)p\delta_{ij}$, the vertical unit vector $\mathbf{e}_z$ pointing upwards, as well as

$$\partial_i p^w = \gamma^w \partial_i h - \gamma_w \mathbf{e}_z \tag{19.28}$$

and the so-called submerged (or buoyant) unit weight $\gamma' := (1-n)(\gamma^s - \gamma^w)$, we obtain from Equations 19.21 and 19.27 the equilibrium equation in terms of effective stresses:

$$\partial_j \sigma'_{ij} + \gamma^w \partial_i h + \gamma' \mathbf{e}_z = 0. \tag{19.29}$$

In geotechnical engineering is distinguished between long-term and short-term strength. The latter refers to the short time after a sudden loading, as long as an overpressure acts in the pore-water. In other words, the undrained case is then considered. In this case, the normal stresses caused by the sudden loading do not increase the frictional shear strength. Consequently, the available strength is only due to pre-existing normal stress. This available strength is called ‘undrained cohesion c_u’. In such problems it is sufficient to consider equilibrium in terms of total stresses, otherwise equilibrium in terms of effective stresses should be considered.

19.6 Consolidation

In the context of water-saturated soil, consolidation means the squeezing out of pore water by the application of external loads. The governing differential equation results from mass conservation of soil grains and pore water, Darcy–Gersevanov's law and a relation for the compressibility of the grain skeleton. We apply the div-operator on Equation 19.16: $\nabla \cdot (\mathbf{v} - n\mathbf{v}^s) = -k\,\nabla \cdot \nabla h = -k\,\Delta h$, for $k = \text{const}$ Δ is the Laplace operator. Herein, we introduce the sum of Equations 19.22 and 19.23 and obtain

$$\nabla \cdot \mathbf{v}^s = \dot{\varepsilon}_v = k\,\Delta h. \tag{19.30}$$

For only one spatial direction z we have $\dot{\varepsilon}_v = \dot{\varepsilon}_z$ and $h = z + u/\gamma^w$, where u is the overpressure, i.e. the pressure exceeding the hydrostatic one. Hence, $\dot{\varepsilon}_z = \frac{k}{\gamma^w} \cdot \Delta u = \frac{k}{\gamma^w} \cdot \frac{\partial^2 u}{\partial z^2}$. The vertical effective stress is $\sigma_z' = \gamma' z + \sigma_0 - u$, hence $\dot{\sigma}_z' = -\dot{u}$. Assuming a constant compressibility (which is rather unrealistic) of the grain skeleton, $\dot{\varepsilon}_z = -a\dot{\sigma}_z' = a\dot{u}$, we finally obtain the consolidation equation for one spatial direction z:

$$\frac{\partial u}{\partial t} = \frac{k}{a\gamma^w} \frac{\partial^2 u}{\partial z^2}, \tag{19.31}$$

which is identical with the Fourier differential equation for heat conduction and diffusion in one spatial dimension. The factor $\dfrac{k}{a\,\gamma_w}$ is also called the consolidation coefficient c_v. The differential equation (Equation 19.31) can be solved by the method of separation of variables or with Laplace transformation. Solutions for common initial and boundary conditions are represented as infinite series [103, 114].

Consider a horizontal layer of water-saturated soil. If it is suddenly loaded by a surface load p_0, then this load is transferred first to the pore water, because the grain skeleton cannot 'feel' any load without being deformed, and compression is delayed by the viscosity of the pore water. Thus, p_0 is only gradually transferred to the grain skeleton, the final settlement s_∞ sets in at $t = \infty$, but a settlement of, say, $0.9s_\infty$ needs a finite time. As with every diffusion process according to Fourier's equation, this time is proportional to the square of a length, in this case the layer thickness, see also Section 18.5.1.

For spatially 3D problems, e.g. for a horizontally infinite soil layer loaded by an impermeable load strip, the pore pressure adjacent to the load strip can initially exceed the value p_0 (so-called Mandel–Cryer effect [71], see Fig. 19.1).

One feels easily tempted to generalise Equation 19.31 to three spatial dimensions simply by replacing $\partial^2 u/\partial^2 z$ by Δu. This is however not justified and leads to an erroneous result.

19.7 Groundwater Flow

The groundwater can flow within the pores of the soil, mainly driven by gravity. Its motion has several consequences for the grain skeleton. Therefore, its mathematical simulation is important.

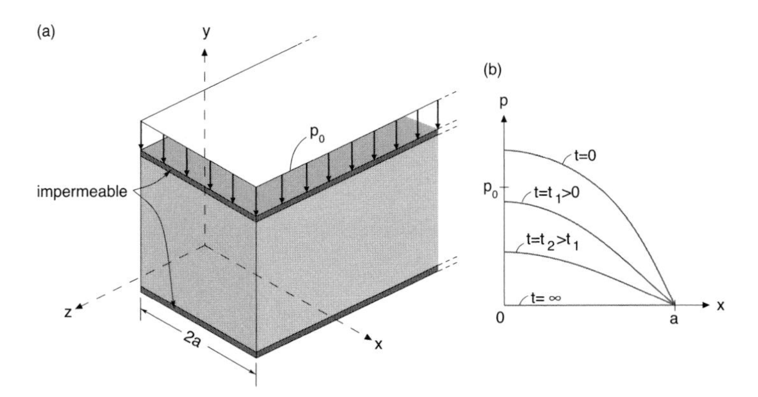

Figure 19.1 (a) Under an impermeable infinite load strip, drainage is only possible in the x-direction. (b) pressure isochrones. Reproduced from [57], courtesy of Springer Nature.

19.7.1 Steady Case

The grain skeleton is often at rest, so that only the velocity of water is considered. Assuming, further, constant porosity n and permeability k, we obtain from Equations 19.15 and 19.23 the differential equation of Laplace: $\Delta h = 0$. For planar problems, $h(x, y)$ depends only on two variables. Solutions of $\Delta h = 0$ are called harmonic functions and can be obtained with conformal mapping (Section 5.3.1). Interestingly, h attains its maximum value at the boundary of the considered domain. Problems with a free water table bear the difficulty that its position is not known a priori and must be searched so that $p = 0$, i.e. $h = z$, holds thereupon.

19.7.2 Unsteady Case

The position of the groundwater table is given by $z = h(x, y, t)$ or $F := h(x, y, t) - z = 0$ [83]. With the components $\frac{\partial x}{\partial t}, \frac{\partial y}{\partial t}, \frac{\partial z}{\partial t}$ of the velocity $\mathbf{v}^w = \frac{1}{n}\mathbf{v}$ and the porosity n, we obtain from $\frac{\mathrm{d}F}{\mathrm{d}t} = \frac{\partial F}{\partial t} + \frac{\partial F}{\partial x}\frac{\partial x}{\partial t} + \frac{\partial F}{\partial y}\frac{\partial y}{\partial t} + \frac{\partial F}{\partial z}\frac{\partial z}{\partial t} = 0$:

$$n\frac{\partial h}{\partial t} + \frac{\partial h}{\partial x}v_x + \frac{\partial h}{\partial y}v_y - v_z = 0, \tag{19.32}$$

where v_x, v_y, v_z are the components of the filter velocity $\mathbf{v}$. For a planar velocity field ($v_y = 0$) follows from this:

$$\frac{v_z}{v_x} = \frac{\partial h}{\partial x} + \frac{n}{v_x}\frac{\partial h}{\partial t}. \tag{19.33}$$

Therefore, the water table can be a streamline only in the steady case ($\frac{\partial h}{\partial t} = 0$).

With the energy head $p/\gamma_w + z$, the filter velocity is given by $\mathbf{v} = \nabla\phi$, where $\phi = -k(p/\gamma_w + z)$ is the so-called velocity potential. At the water table we have $p = 0$, therefore $\phi(x, y, h, t) = -kh$. Now the assumption of Dupuit is used: The horizontal velocity components v_x and v_y do not depend on z. It then follows from $\operatorname{div}\mathbf{v} = 0$:

$$\frac{\partial v_z}{\partial z} = -\left(\frac{\partial v_x}{\partial x} + \frac{\partial v_y}{\partial y}\right) \rightsquigarrow v_z(z) - v_z(z = 0) = -\int_0^z \left(\frac{\partial v_x}{\partial x} + \frac{\partial v_y}{\partial y}\right)\mathrm{d}z. \tag{19.34}$$

With $v_z := v_z(z)$ and $v_{z0} := v_z(z = 0)$ this results in a linear relationship between v_z and z:

$$v_z = -\left(\frac{\partial v_x}{\partial x} + \frac{\partial v_y}{\partial y}\right) z + v_{z0}. \tag{19.35}$$

Since Equation 19.35 is based on an averaging over depth, v_{z0} can include contributions due to a vertical injection. Now we insert $v_x = -k\frac{\partial h}{\partial x}$ and $v_y = -k\frac{\partial h}{\partial y}$ into Equation 19.32:

$$n\frac{\partial h}{\partial t} - k\left[\left(\frac{\partial h}{\partial x}\right)^2 + \left(\frac{\partial h}{\partial y}\right)^2\right] - v_z = 0 \tag{19.36}$$

and then we express v_z with Equation 19.35 and $z = h$:

$$n\frac{\partial h}{\partial t} - k\left[\left(\frac{\partial h}{\partial x}\right)^2 + \left(\frac{\partial h}{\partial y}\right)^2\right] - kh\left(\frac{\partial^2 h}{\partial x^2} + \frac{\partial^2 h}{\partial y^2}\right) - v_{z0} = 0. \tag{19.37}$$

Considering $\frac{\partial}{\partial x}\left(h\frac{\partial h}{\partial x}\right) = \left(\frac{\partial h}{\partial x}\right)^2 + h\frac{\partial^2 h}{\partial x^2}$ leads finally to the differential equation of Boussinesq:

$$\frac{\partial h}{\partial t} = \frac{k}{n}\left[\frac{\partial}{\partial x}\left(h\frac{\partial h}{\partial x}\right) + \frac{\partial}{\partial y}\left(h\frac{\partial h}{\partial y}\right)\right] + \frac{v_{z0}}{n} \tag{19.38}$$

or

$$\frac{\partial h}{\partial t} = \frac{k}{n}\operatorname{div}(h\nabla h) + \frac{v_{z0}}{n}, \tag{19.39}$$

where div and ∇ are to be expressed with the horizontal coordinates x and y or r and θ. With $h\nabla h = \frac{1}{2}\nabla h^2$ we get the so-called porous media equation

$$\frac{\partial h}{\partial t} = \frac{k}{2n}\operatorname{div}(\nabla h^2) + \frac{v_{z0}}{n} = \frac{k}{2n}\Delta h^2 + \frac{v_{z0}}{n}, \tag{19.40}$$

where Δ is the Laplace operator. The general form of this partial differential equation is $\partial h/\partial t = c\Delta h^m$. It applies, for example, to the flow of gas in a porous medium. Darcy's law is: $v = -k'\nabla p$, with $p = p_0\rho^\gamma$. Inserting into the equation of the mass balance, $\rho_t + \operatorname{div}(\rho v) = 0$, returns the equation $\rho_t = c\Delta\rho^{\gamma+1}$.

Remark Electro-osmosis: The flow of groundwater is not only induced by pressure gradients, but also by gradients of the electric potential (i.e. through electric fields). This phenomenon is called electro-osmosis and is based on the fact that cations in the groundwater are attracted to the cathode, and as they move there, they carry away water molecules due to their dipole moment. With electro-osmosis Darcy's law extends to $v = ki + k_e E$. Here, E is the electric field strength and k_e the so-called electroosmotic permeability For almost all soils it equals approximately $5 \cdot 10^{-5}$ cm^2/V·s.

Remark Anisotropic permeability: The permeability is anisotropic if it does not have the same value in all directions. Darcy's law is then $v_i = -k_{ij}\frac{\partial h}{\partial x_j}$, and k_{ij} is the permeability tensor. For example, for earth dams, the layer-by-layer installation and compaction implies that the horizontal permeability k_h is greater than the vertical one k_v. It should be added however that the ratio k_h/k_v can hardly be measured.

Laminar flow in rock joints can be regarded as Couette flow. If the joints are parallel with a spacing s and with the opening width b, then the global permeability in the direction of the joints is $k = \gamma_w b^3/(12s\mu)$, e.g. for $s = 1$ m and $b = 1$ mm it is $k \approx 10^{-3}$ m/s. The cubic law ($k \propto b^3$) is valid up to aperture widths of 10 μm. If this permeability is homogenised (smeared) over the rock mass, the following tensorial relation is obtained

$$v_i = -k_{ij}\frac{\partial h}{\partial x_j} \quad \text{with} \quad k_{ij} = \frac{\rho^w g}{\mu}\frac{b^3}{12s}n_i n_j, \tag{19.41}$$

where n_i is the unit vector normal to the joints.

19.7.3 Transport of Pollutants in Groundwater

Pollutants and other substances can move and propagate in the groundwater. Three different mechanisms can drive this motion:

Convection: The essential transport mechanism is convection (advection), i.e. the pollutants are carried along by the flowing groundwater. Convection is given by the equation

$$\partial c/\partial t + \nabla \cdot (c\mathbf{v}) = 0,$$

which describes the mass conservation of the pollutant, c is its concentration (see also Section 13.3).

Diffusion is driven by the thermal excitation of molecules and leads to the so-called Brownian molecular motion. It is described by the differential equation

$$\mathbf{q} = -nD\,\nabla c,$$

where $\mathbf{q}$ is the flux (i.e. the quantity of the pollutant that flows through a unit area in a unit time) and D is the diffusion coefficient. Molecular diffusion causes transport even in stagnant groundwater. Compared with diffusion in clean water (diffusion coefficient D_0), for diffusion in the pore space the coefficient $D = D_0/\lambda^2$ applies, where λ is the so-called tortuosity [67]. With respect to any two points in groundwater with distance a, we have $\lambda = b/a$, where b is the length of the shortest connecting curve in the pore space.

Hydrodynamic dispersion: This is based on the fact that the water velocity $\mathbf{v}$ or $\mathbf{v}^w$ is only a temporal and spatial mean value. The actual velocity fluctuates around this value due to the granular structure of the soil (Fig. 19.2). The flux $\mathbf{q}_h$ due to hydrodynamic dispersion is proportional to the concentration gradient ∇c:

$$\mathbf{q}_h = -\mathbf{D}_h\,\nabla c.$$

$\mathbf{D}_h$ is the so-called hydrodynamic dispersion tensor.

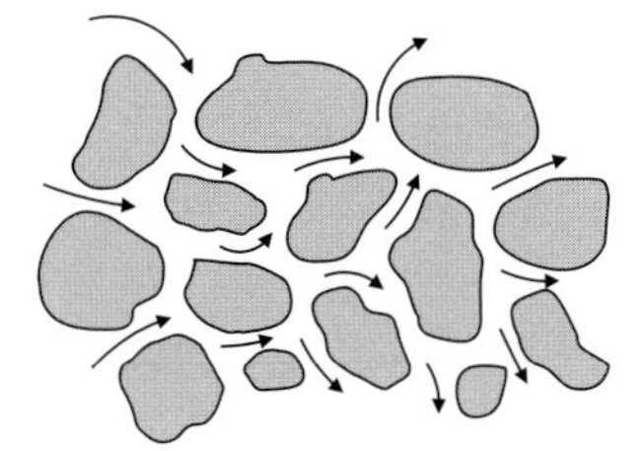

Figure 19.2 Real velocity of water within a grain skeleton. Reproduced from [57], courtesy of Springer Nature.

19.8 Unsaturated Soil: An Exkurs to Physical Chemistry

In unsaturated soil, the pore volume is filled partly with water and partly with air. Although the liquid and gaseous phases are in contact, the pressures p_w and p_a prevailing in them are not equal. Their difference, the so-called suction, is caused by capillarity and osmosis. The question is how p_w and p_a influence the mechanical behaviour of the grain structure. The considerations linked to unsaturated soil draw heavily on physical chemistry, and the experimental techniques to measure suction are elaborate. However, the results obtained so far from the research on the mechanical behaviour of unsaturated soil are still doubtful.

19.8.1 Capillary Suction

Capillarity is caused by surface effects, i.e. phenomena at the surface of the individual phases. In liquids and cohesive solids, the particles at the phase boundary experience a non-vanishing resultant of the molecular attraction forces (Fig. 19.3), so the boundary plays a prominent role energetically.

γ is the surface tension of the considered liquid with respect to the adjacent phase. If the surface is understood as a thin membrane, then γ is the membrane tensile stress. If the interface is part of a sphere with radius r, then Laplace's equation gives the pressure difference Δp^{cap} between the two phases:

$$\Delta p^{cap} = \frac{2\gamma}{r}. \tag{19.42}$$

The overpressure Δp^{cap} prevails in the convex phase, such that there is a relative negative pressure in the concave phase. For example, in a water droplet with diameter of 1 μm there is an overpressure of approximately 3 bar. For unsaturated soil, γ is the surface tension for the interfaces between air and water. Equation 19.42 implies that water pressure and air pressure in unsaturated soil are neither equal nor independent of each other but differ by the amount Δp^{cap}. If the air pressure p_a is equal to atmospheric pressure, then the water pressure is $p_w = 1$ atm $- \Delta p^{cap}$, i.e. a negative

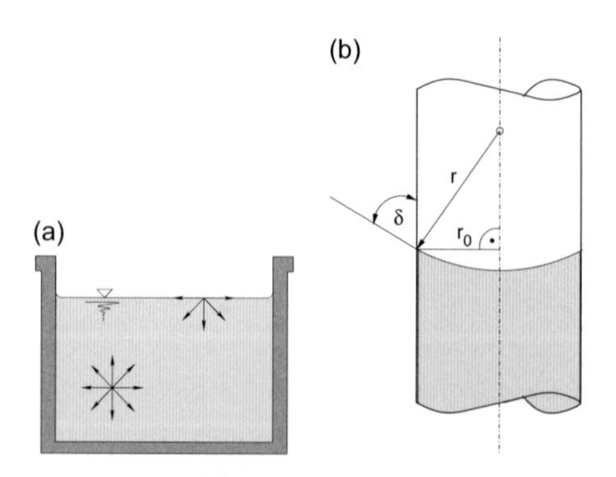

Figure 19.3 (a) Molecular attractions in the interior and at the surface of the liquid phase. (b) Meniscus of a fluid in a capilar tube. Reproduced from [57], courtesy of Springer Nature.

pressure prevails in the water. Therefore, $\Delta p^{cap} := p_a - p_w$ is the suction of the grain matrix.

Common units of pressure:
cm or m water column, Pa = N/m^2, psi = lb/in^2, pF = ln(cm water column)
100 kPa = 1 bar = 14.5 psi = pF 3 $\approx$ 10 m water column.
At mean sea level: 1 atm = 101.3 kPa $\approx$ 1 bar.

Actually, a liquid evaporates when the vapour pressure is reached (cavitation), so that high tensile stresses should not occur in liquids. However, cavitation requires the presence of evaporation nuclei, the probability of which decreases with decreasing volume. Thus, in fluids within small volumina, tensile stresses of up to the order of magnitude of the so-called molecular tensile strength can occur. The latter is likely to be of the order of 1000 bar for water at 20°C.

Remark For surfaces with double curvature, the term $2/r$ in Equation 19.42 should be replaced by the sum of the maximum and minimum curvatures. Capillary surfaces are surfaces of constant mean curvature and attain special consideration in differential geometry.

For capillary tubes with a circular cross section, r results from the radius r_0 of the tube and the contact angle δ (Fig. 19.3). The latter represents a material property and results from the surface tensions $\gamma_{aw}, \gamma_{as}, \gamma_{ws}$ (the indices indicate the respective contacting phases) according to Young's equation [93]:

$$\cos\delta = \frac{\gamma_{as} - \gamma_{ws}}{\gamma_{aw}}.$$

As can be seen from Fig. 19.3, r is obtained from:

$$r = r_0/\cos\delta.$$

For soil, r is simply an average value reflecting the geometrical and material properties of the grain structure (with respect to water and air).

19.8.2 Osmotic Suction

If a saline pore water is in contact with a reservoir containing water of lower salt concentration, the cleaner water is drawn into the pore space. The associated suction is called osmotic. The dissolved salts gradually diffuse into the 'clean' water unless they are prevented from doing so, for example, by electric fields on the surfaces of the solid particles.

The osmotic pressure p^{osm} is given by van't Hoff's law:

$$p^{osm} \cdot V = n \cdot R \cdot T.$$

Herein, n is the number of moles in the solution. The osmotic pressure is therefore equal to the pressure which the dissolved substance would exert as a gas with the same values of V and T. A higher salinity of the pore water, as compared with the water outside the grain matrix, gives rise to the osmotic suction Δp^{osm} such that in the literature of unsaturated soil the total suction Δp^{tot} is often considered as composed

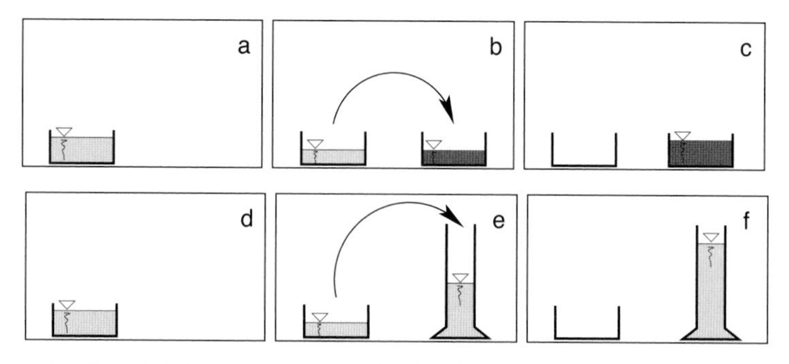

Figure 19.4 Water extraction through air. (a–c) osmotic suction, (d–f) capillary suction. Reproduced from [57], courtesy of Springer Nature.

of the capillary *and* osmotic parts $\Delta p^{tot} = \Delta p^{cap} + \Delta p^{osm}$. However, the role of osmotic pressure in the mechanics of unsaturated soil is yet unclear, and therefore, it will not be considered further here.

19.8.3 Water Transport through Air

Osmotic and capillary suction can draw in water through the air. Fig. 19.4 showns a shell filled with water in an enclosed space. Thermodynamic equilibrium means that the amount of water evaporating is in equilibrium with the amount of water condensing. Now we put a second bowl into the vessel, which contains a concentrated salt solution. The dissolved salt lowers the vapour pressure, i.e. more water molecules condense into the salt solution than evaporate (because of the osmotic suction, the salt solution is 'eager' for water). In thermodynamic equilibrium, a lower vapour pressure is reached, whereby the water evaporates from the first shell and moves into the second one (principle of the exsiccator). The vapour pressure of the salty water p_v is smaller than the vapour pressure of pure water p_v^0. According to Raoult's law, the ratio p_v/p_v^0 is equal to the mole fraction ν_{H_2O} of the water.

Mole fraction: number of particles n_i of a given substance i in relation to the total number of particles n: $\nu = n_i/n$. 1 mole of each substance contains L particles, where L is the Loschmidt number ($L = 6.025 \cdot 10^{23}$).

1 mol water has the mass $m_{H_2O} = 18.02$ g, $m_{O_2} = 32.0$ g, $m_{air} = 28.95$ g, $m_{CO_2} = 44.01$ g.

If one mixes the substance i (mass M_i) with the substance j (mass M_j), then the mole fraction of the substance i is

$$\nu_i = \frac{M_i/m_i}{(M_i/m_i + M_j/m_j)}.$$

The volume of 1 mol is called **molar volume.**

Just like a salt solution, a water–air interface curved by surface tension can lower the vapour pressure (water molecules have a harder time evaporating out of a convexly curved water–air interface). A capillary tube or a porous substance thus taps the water from the shell (as shown in Fig. 19.4d–f). In both cases, the evaporation of the water in the shell lowers the temperature. This is the principle of the so-called psychrometers, which measure suction via the coldness of evaporation. The theoretical basis for this is the Kelvin equation, which relates the ratio of the reduced vapour pressure p_v due to suction to the vapour pressure p_v^0 (without suction) with the suction tension Δp^{cap}:

$$\Delta p^{cap} = \frac{1}{v} \ln \frac{p_v^0}{p_v}, \quad \text{for water at } 20^\circ\text{C:} \quad \Delta p^{cap}(\text{kN/m}^2) = -135.055 \ln \frac{p_v}{p_v^0} \tag{19.43}$$

Hierin, v is the molar volume of water, R is the gas constant, T is the absolute temperature and p_v/p_v^0 is also called the relative humidity.

19.8.4 Air Transport through Water

Water and air can migrate via diffusion into each other. Consider, for example, a water-saturated soil sample that is laterally confined. If it is exposed to air of increased pressure (not exceeding the air entry value) at the one side, diffusive air transport through the pore water will set in. If a gas is in contact with a liquid, a part of the gas penetrates the liquid in dissolved form. The (vapour) pressure of the dissolved gas p is proportional to its mole fraction ν (Henry's law): $p = K\nu$. The temperature-dependent material constant K is called the Henry constant [4].

19.8.5 Airflow in the Pores

For airflow through an unsaturated soil with continuous air channels, Darcy's law applies: $\mathbf{v}_a = \text{const} \cdot \nabla p_a$. Considering the air compressibility ($p_a = \kappa \varrho_a^\gamma$ for adiabatic compression), mass balance of air leads to the so-called *porous media equation*

$$\frac{\partial \varrho_a}{\partial t} - \text{const} \cdot \Delta(\varrho_a^{\gamma+1}) = 0.$$

19.8.6 Water Conduction in Unsaturated Soil

Darcy's law $v = -k\nabla h$ applies here also; the filter velocity v represents the volumetric water flux [27]. However, the usual expression $h = \frac{p}{\gamma^w} + z$ is not applicable in cases where a low saturation obstructs a continuous water body. To consider the water velocity due to a gradient of suction we set

$$h = \frac{p^{cap}}{\gamma^w}. \tag{19.44}$$

Often, reference is made to the so-called volumetric water content $\theta := V_w/V$. Obviously, $\theta = nS$, where n is the porosity and S the degree of saturation. k depends on θ or (for $n = \text{const}$) on S and, thus, it varies in space and with time. Therefore, its experimental determination is very elaborate. With $\mathbf{v} = -k\nabla h$ and the mass balance for water div $\mathbf{v} = -\partial\theta/\partial t$ it follows:

$$\nabla \cdot (k\nabla h) = \frac{\partial \theta}{\partial t}. \tag{19.45}$$

19.8.7 Filter

Particularly subtle in the case of saturated and unsaturated soils is the question of the boundary conditions: If the total stress is prescribed on the boundary, then it is unclear how it is partitioned between the grain skeleton and the pore phases. If the air pressure is given, then the pore-water pressure is also given (offset by the amount Δp^{cap} according to Laplace). How can we apply a pressure specifically on the air or the water? Porous discs are only permeable to pore fluids (air and water) but not to grains. So, they are a kinematic boundary condition for the grains, whereas they impose a static boundary condition to the pore fluid. In addition, porous discs are only permeable to a limited extent: If a water-saturated porous disc is adjacent to a coarser unsaturated grain skeleton or to air, an excess air pressure will not displace the water from the porous disc as long as the overpressure is less than the so-called air entry value (or bubbling pressure). This can be seen with Laplace's equation: If the pores of the filter have diameter r, then the air entry value, i.e. the required air overpressure, is $2\gamma/r$. Conversely, an overpressure of the water will not cause the air to flow out, as long as it is less than this threshold value. This is the principle of the *capillary barrier*, which is used, for example, to seal the surface of landfills.

19.8.8 Soil–Water Characteristic Curve

The so-called soil–water characteristic curve (SWCC) $s(S)$ indicates the suction $s := \Delta p = p_a - p_w$ as a function of the degree of saturation S, see Fig. 19.5. The behaviour is hysteretic so that the curve is divided in two branches. The steep increase of s in the drying curve by the value p_e (or s_e) is due to the fact that air can pass through a pore with a diameter of $2r$ only when the air overpressure reaches the air entry value $s_e = 2\gamma/r$. Expelling the water from the pore space (desorption, drying, dehumidification) and the drawing in of the water into the pore

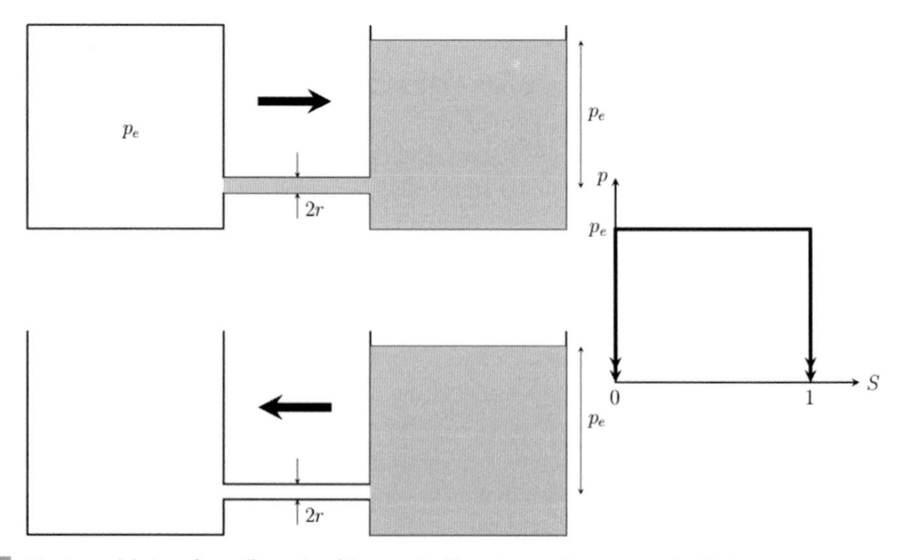

Figure 19.5 Wetting and drying of a capillary tube of diameter $2r$. The action sets in as soon as the driving pressure reaches the air entry value p_e.

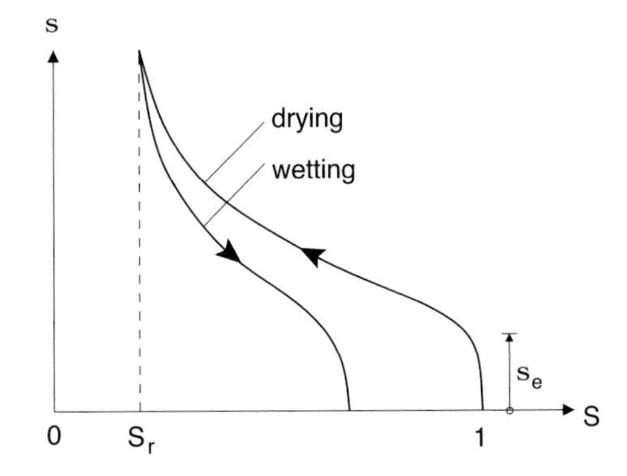

Figure 19.6 SWCC, drying and wetting branches, S_r = residual degree of saturation, s_e = air entry value. Note that this curve represents the behaviour of a sample and is, strictly speaking, not a material property. Reproduced from [57], courtesy of Springer Nature.

space (absorption, wetting, humidification) occur as soon as the pressure reaches the value s_e, see Fig. 19.5 for an idealised pore and Fig. 19.6 for a real soil. The usual experimental procedures for determining the SWCC of a grain matrix are based on the displacement of either the air or the water from the pore space by applying a pressure to the sample boundary.

19.8.9 Effective Stresses in Unsaturated Soil

By averaging over the pore area in an arbitrary cross section and using Delesse's theorem one obtains the pore pressure averaged over the pore volume:

$$p = (1 - S)p_a + Sp_w,$$

where p_a is the air pressure and p_w is the water pressure. If one adheres to the definition of effective stress for water-saturated soils, one obtains (for the sake of simplicity no indices are written for the tensor components):

$$\sigma' = \sigma - p = \sigma - (1 - S)p_a - Sp_w = \sigma - p_a + S(p_a - p_w). \tag{19.46}$$

Compressive normal stresses are positive here, correspondingly suction is negative. $s = p_a - p_w$ is the total suction. The part $\sigma - p_a$ is called 'net stress'. For $S = 1$ Equation 19.46 reduces to the conventional definition of the effective stress. This is also the case for dry soil with $S = 0$. For intermediate values, $0 < S < 1$, the validity of Equation 19.46 is still unclear despite many publications in the field (see, e.g., [41]).

20 Computing in Soil Mechanics

20.1 Pitfalls of Computing

Calculations play an important role in civil engineering. On a bridge, in a plane or in a skyscraper, we feel safe because it has been calculated. However, there are also computationally intensive disciplines such as meteorology, where the reliability of calculation results is modest. This is also the case with soil mechanics calculations in geotechnical engineering. The main reasons for this are probably the non-linear and 3D problems, the strong scatter, and thus deficient input data, and the physical principles (in particular the constitutive law of soil) that have not yet been completely mastered. In this field of uncertainties, a number of misconceptions are established, such as that

- the quality of the results increases with the complexity of the calculations,
- the quality of the results increases with the quantity of input data. Further, there is an expectation that data, much more data, can reveal physical laws, if properly processed ('data mining') and
- computer results are altogether reliable.

Terzaghi [104] warns not to expect too much accuracy of calculation results:

> ...the illusion that everything connected with engineering should and can be computed .... In soil mechanics the accuracy of computed results never exceeds that of a crude estimate, and the principal function of theory consists in teaching us what and how to observe ...

As an example, it should be noted that in the early days of tunnelling, the lining of tunnels was empirically designed. Nowadays it has to be calculated, no matter that we hardly know the thrust exerted by the adjacent rock. Usually, a thrust is 'invented' so as to justify a reasonable lining.

The Cartesian belief in the computability of everything and also commercial and publicity skills are behind some calculation results. Truesdell [111] cites H.R. Post:

> The Computer is certainly the perfect instrument for inevitable research, that is, research which is certain to deliver some 'results', right or wrong or meaningless, on any problem proposed.

The power of modern computers has even led some researchers to look at every single grain of soil, whether as a sphere or even in a more natural form. Taking account of such a large number of grains is expected to yield valuable results for many problems. Looking at tiny grains might reveal some aspects, but one can also get lost in the detail.

20.2 Problems with Geotechnical Engineering Computations

20.2.1 Scatter of Soil Properties

In contrast to, e.g. steel, which is an artificially produced material with controlled properties, soil deposits are created by geological processes and exhibit a spatial scatter. Several attempts to detect rules of this scatter (see, e.g., [39]), such as autocorrelation, covariance lengths, variogramms and kriging, cannot be promising because soil deposits do not have the 'mixing property': an initial irregularity, e.g. due to a boulder, a cavity, a tree trunk, a fault, does not smooth out with time. This is to be contrasted with gases, where diffusion smooths out the fluctuations of a gas concentration.

Some scientists cast doubt on all numerical computations in geotechnics because of the spatial scatter of soil properties. This is however not justified, as this scatter can be – in principle – taken into account by cautious assumptions. In the field of constitutive relations, soil can be assumed as a homogeneous material, samples of (at least initially) homogeneous soil can be investigated in the laboratory, and there are ever-improving constitutive laws for soil. In this respect, soil mechanics is to be contrasted with rock mechanics, where computations have a much more difficult stand. With rock, a distinction must be made between hard and 'competent' rock (e.g. Scandinavian granite) on the one hand, and jointed or weathered rock of reduced strength on the other hand, where most of the technical problems arise. In the first case, calculations are possible (e.g. according to the theory of elasticity) but rather superfluous, whereas they are hardly possible in the second case because of the pronounced effect of geometric scale (jointed rock is not a 'simple' material). To put it somewhat casually: computations in rock mechanics are only possible when they are not needed. Hence, empirical relations (e.g. based on the so-called rock quality designation or other similar indices) seem to be more purposeful in rock mechanics than investigations based on mechanical analysis. As a consequence, the name 'rock engineering' appears more appropriate than 'rock mechanics'.

20.2.2 Initial Stress Field

Soil stiffness depends on stress. This is to be contrasted with linear elasticity, where Young's modulus E is constant. The resulting difficulty for soil is enormous, as it requires us to know the initial stress field for every computation. The theory of plasticity helped sweep this problem under the carpet, because it assumes that soil behaves initially as an elastic material.

For the case of a sedimented and non-preloaded soil halfspace with a horizontal surface, the initial stress field can be assumed as geostatic, i.e. $\sigma_z = \gamma z, \sigma_x = \sigma_y = K_0 \sigma_z$. For other cases, the geologic history should be – strictly speaking – numerically simulated starting from a geostatic stress field.

20.2.3 Boundary Conditions

For almost every geotechnical problem, the area of interest is the so-called halfspace, i.e. an infinitely extended body. This is to be contrasted with the structural analysis of, e.g. buildings, where we deal with finite bodies such as a beam. The infinite extend of the halfspace poses the question of how big a section of it we should consider. It appears reasonable to consider a sufficiently large section, so that the arbitrarily set boundaries have an as small as possible influence on the results. A difficulty arises in dynamical problems, where waves are reflected at the boundaries unless special absorbing dashpots are provided for.

Of particular interest in geotechnical engineering are the so-called *soil-structure-interaction (SSI)* problems. Here, the interaction of soil with a structural element such as a pile or a foundation beam is of interest. One looks at the deflection y of a beam or pile as a function of the longitudinal coordinate x. Clearly, the curve $y(x)$ depends upon the load $p(x)$ acting upon it: $y(x) = \mathcal{F}[p(x)]$. The difficulty arises from the fact that the functional $\mathcal{F}$ is different for the beam (or pile) and for the soil. The solution $y(x)$ has to be sought iteratively so as to fulfil both functional relations, for the beam (or pile) and for the soil. For the beam, the relation between the functions $y(x)$ and $p(x)$ is given by the differential equation $EJy^{(4)} = -p(x)$ and the individual loads acting upon it. For soil, this relation is more complex and has to be found numerically.

20.2.4 Improper Quantities

We should be cautious in operating with improper variables. In classical physics (but not in quantum mechanics) a physical quantity has a unique value in every state of a system [22]. The safety factor and the subgrade reaction modulus are examples of non-physical quantities. A beam has a weight, a temperature, a velocity, but not *a* safety factor. There are many different conventions for safety factors, but these conventions are inherently arbitrary and, thus, non-unique. Hence, the statement 'This beam has *the* safety factor X' is not a scientific statement because it cannot be verified or (according to Popper) falsified. In particular, safety cannot be measured. Of course, this is awkward for any measure to strengthen a construction which seems to be endangered, e.g. to strengthen a slope by anchors. In many cases we can and we should install anchors, but it is not possible to quantify the resulting benefit.

The so-called subgrade reaction modulus is often used to simplify problems of soil structure interaction. It is assumed that the soil reacts at a location x with a pressure p that depends *only* on the displacement y at this location. In other words, the soil is assumed as a collection of infinitely thin (not necessarily linear) springs, that are independent of each other. The assumption $p = k \cdot y, k = \text{const}$, facilitates enormously the problem stated in Section 20.2.3, as it leads to the linear differential equation $EJy^{(4)} = -ky$, which has computationally intensive but manageable solutions. However, there is no way to measure the subgrade reaction modulus k, because it is based on a concept that is physically unsound.

Also, Young's modulus is an improper quantity when referring to the behaviour of soil, because it refers to Hooke's law, which is invalid for soil (with the exception of small strain cycles considered by soil dynamics).

20.3 Limits of Continuum Mechanics

The main advantage of continuum mechanics is the applicability of (infinitesimal) calculus to solve problems, 'even in such complicated cases, where without such help even the genius becomes impotent', according to Gauss [99]. Continuum mechanics considers mainly so-called simple materials, but in some cases this assumption is not justified, so researchers resort either to so-called higher continua or to consideration of the individual grains.

There are cases that can be hardly predicted by calculations. When we toss a coin, we cannot calculate which side it falls on. Similarly, we can calculate at which load a compressed rod will buckle, but we can hardly predict to which side it will buckle. A similar situation is encountered in geotechnical engineering and is vividly illustrated by numerical calculations. A computer, which in the end does nothing else than solve systems of linear equations, stops execution of the program and reports that there is no longer a unique solution. This signifies that the determinant of the system vanishes and implies the existence of more than one solution. Clearly, the computer is incapable of deciding which one to follow further. This ambiguity often signifies that nature is also undecided about which option to pursue further. Of course, a compressed rod in the end buckles in one direction, which however depends on minute details that are difficult to capture. In geotechnical engineering, loss of uniqueness (so-called bifurcation) signifies a sort of collapse, and a continuation of the computation makes little sense, as the further development depends on barely detectable details. Nevertheless, many researchers try to follow this route. To do this, they introduce additional assumptions, and this is called regularisation. The sense of such analyses remains often questionable.

Some authors miss structural details of the material that are lost in the transition to the infinitely small in continuum mechanics: 'When using the differential formulation, a length scale must be introduced into the material description of a strain-softening modelling. This need has been here justified on the basis of the geometrical information which has been lost in performing the limit process' [23]. The missed detail is essentially the grain size. It is decisive for the thickness of shear bands, which are linked with softening. Much effort is spent trying to predict their thickness. Apart from the questionable usefulness of this information, it should be added that the few calculations on this are based on so many assumptions that it would be more direct to assume the result.

20.4 Quality of Numerical Results

In view of the confusing variety of constitutive equations and numerical solution methods of initial boundary value problems (IBVPs) and their parameters (determination of the material parameters, discretisation, termination criteria for iterations, etc.) we should consider results of numerical simulations critically, but not condemn them. The numerical simulation manifests the quest for understanding the underlying relationships. Critical reflection will help to select better procedures. However, this requires that simulation results are properly documented. The criterion for completeness of documentation is that it allows a third party to understand and repeat the numerical simulation.

The working group 'Numerical Methods in Geotechnics' of the DGGT (German Society for Geotechnical Engineering) has carried out an interesting investigation [94]: a tunnel with prescribed cross-section and boundary conditions, prescribed constitutive equation and prescribed material constants was calculated by different participants. While the obtained surface settlements showed no significant scatter, the normal force and the bending moments in the lining varied by up to 300%!

For a retaining wall near Hochstetten, geotechnical engineers were invited to predict the distribution of earth pressure, bending moments and displacements based on test results on the soil [119]. The scatter of the predictions is enormous (Fig. 20.1).

Has this situation improved since the early 1990s? A similar large-scale experiment was recently carried out in Australia. The predictions of the force-displacement behaviour of a single foundation in clay exhibited a large scatter [21].

Note that pressure distributions, for example on foundation plates or tunnel linings, can hardly be measured. Thus, related computations are hardly ever validated by comparing them with measurements.

Often, the results of a calculation are checked against the results of other calculations. This may show the deviation and the scatter across them, but a veritable check (validation) can only be obtained against measurements. In this respect is interesting the second thesis on Feuerbach by Marx and Engels: 'The question whether objective truth can be attributed to human thinking is not a question of theory but is a practical question.'

20.5 Coping with Uncertain Predictions

Many disciplines exist and even flourish despite a fundamental lack of knowledge in their fields and the resulting uncertainty. Think of investment managers, physicians, politicians and others. These branches have developed strategies and techniques to deal with incomplete knowledge. In geotechnical engineering the following techniques are applied:

1. Increased safety margins for the however designed constructions.
2. Cautious building by application of the observational method (Section 20.5.1).
3. Reliance on authorities. Instead of independently assessing a situation, an external judgement is sought. One resorts to other people's (pretended) knowledge. Clearly, this is not a truly scientific method, as the results cannot be checked on a rational basis. Authorities are called in the following ways:

    1. Computer programs. They are based on some numerical method and on some constitutive law. Usually, both are inaccessible to the users and, hence, they are

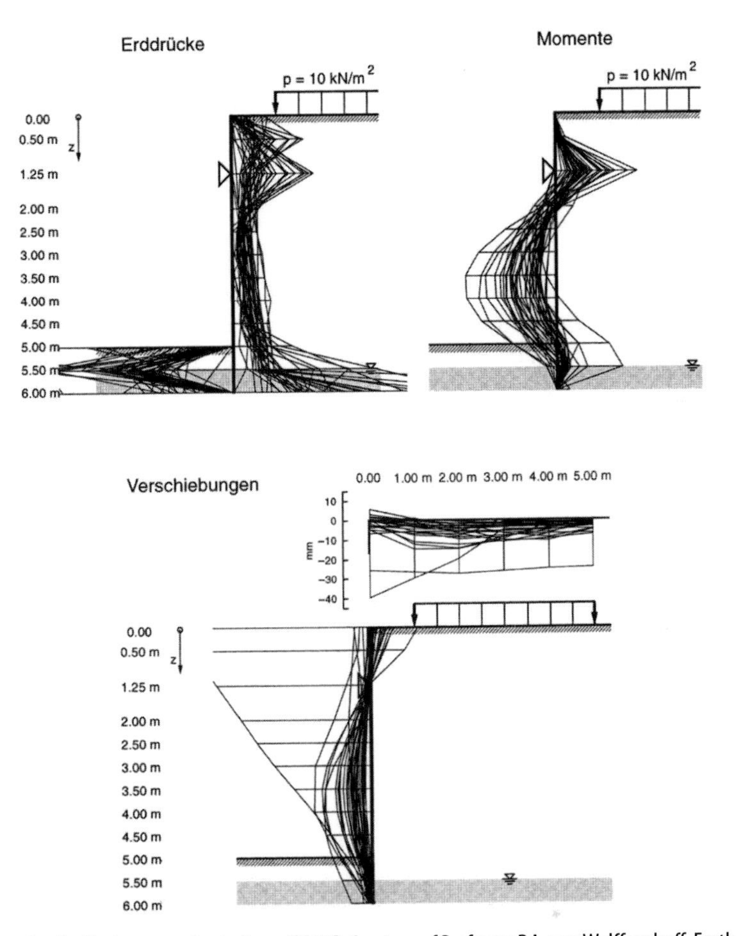

Figure 20.1 Predictions for the Hochstetten sheet pile wall [119]. Courtesy of Professor P.A. von Wolffersdorff. Earth pressures, bending moments and displacements.

applied as black boxes. Their (scarce) validations show a rather unsatisfactory accuracy of the results, which exhibit a large scatter.

2. Codes of practice. Usually, these are not intended to be tractable. Their increasing multiplicity and complexity do not support understanding nor applicability.
3. Experts. Especially in rock mechanics, experts are called in to answer open questions based on their intuition. Experts learn by long training (similar to artificial intelligence), but they are mostly unwilling or incapable of editing their knowledge or to share it with others.

20.5.1 The Observational Method

As calculations yield uncertain results, we should adapt them during construction *(learn-as-you-go)*. The observational method [79] is a gradual approximation to the actual conditions and consists of the following steps:

1. Site investigation.
2. Determine which are the most probable and which are the least favourable subsurface conditions to be expected (here geology plays a major role).
3. Design, based on the most likely conditions.
4. Select the variables to be observed during the construction process. Calculate their expected values. This should be done for both the most probable and the least favourable conditions.
   Hence, the observational method does not make calculations superfluous (as is commonly assumed).
5. A priori determine the actions to be taken when the observed values reach certain intervention limits.
6. Measure (observe) and possibly adjust design.

The observational method is based on the idea that collapse is announced by increased deformation. Hence, it is meaningless for so-called brittle systems, where failure sets in without announcement.

20.5.2 Soil as Gracious Material

It may seem surprising that, in spite of scattering calculation results and the confusion created by the many codes, there are not more cases of damage in geotechnical engineering. On the one hand, the reason lies in the usual over-dimensioning. On the other hand, however, it is due to the quality of the soil, which usually forgives the yielding of a supporting structure. This is because earth pressure (in contrast to water pressure) is displacement dependent: when a retaining wall yields, the earth pressure is reduced. This has something to do with the mobilisation of shear stresses and the effect of vaulting and is exploited in geotechnical engineering, sometimes consciously, sometimes unconsciously. In any case, the conditions are similar to those exploited by Asian martial arts: yielding relieves pressure. In contrast, water pressure is independent of displacement, therefore easy to calculate, but correspondingly unforgiving. If the interaction of the soil with a retaining wall is characterised by principles of forgiveness and kindness, water mercilessly uses any yielding and thus proves to be the worst enemy of the geotechnical engineer. The superiority of water is well known, see Lao Tse:

> Nothing in the world is as soft and yielding as water.
> And yet it conquers the hard and strong.

In fact, most damage in geotechnical engineering is due to the action of water.

21 Outlook

21.1 Open Questions

Given the central importance of constitutive modelling in theoretical soil mechanics, the following points should be addressed.

At present, the interest of soil mechanics in constitutive modelling is reduced, mainly because of the complexity of the known models. Most of them are hidden as black boxes in computer codes, and only a few specialists deal with them. The involved scientific community is highly fragmented, the complexity of the models inhibits a fruitful scientific exchange and progress. This is a pity in view of the many still unsolved problems. To list some of them:

- How can we link constitutive models with the grain characteristics (grain sizes and grain size distribution, shape of grains, etc.) of the individual soils? For example, how can we predict the mechanical behaviour of a soil that is obtained by mixing two other soils?
- The constitutive modelling, as presented in this book, describes granulates as grain skeletons where the grains are always in contact with each other. However, granulates are also encountered in other phases, e.g. as fast flowing debris or even as granular gases. See the so-called dense phase and dilute phase transports of granulates in industry. How can we describe the related phase transitions?
- Whereas soil mechanics is mainly concerned with the irreversible behaviour of grain skeletons, soil dynamics operates with linear elasticity. There is a veritable gap between these two disciplines, which both serve geotechnical engineering but still use different methods and paradigms. It would be beneficial to establish a link between these branches. A similar gap exists between soil mechanics and physics, which increasingly considers granulates and applies methods of molecular dynamics, ignoring though the approaches and findings of engineers.
- Granular materials are the main representative of *soft solids*. Given that rock can also be considered as a granular material, a good command over the mechanics of granulates would make it possible to realistically simulate geologic processes, since this is already the case with sandbox models for the physical simulation of geologic processes.

References

[1] Abedi, S., Rechenmacher, A.L. & Orlando, A.D., Vortex formation and dissolution in sheared sands. *Granular Matter* (2012), 14:695–705.

[2] Achenbach, J.D., *Wave Propagation in Elastic Solids*, New York: American Elsevier, 1973.

[3] Amorosi, A. & Rampello, S., An experimental investigation into the mechanical behaviour of a structured stiff clay. *Géotechnique* (2007), 57(2):153–166.

[4] Atkins, P.W., *Physikalische Chemie*, Hoboken, NJ: Verlag Chemie, 1990.

[5] Bathaeian, I., Meshfree simulation of problems in soil mechanics. PhD thesis, Universität Innsbruck, 2018.

[6] Becker, B. & Bürger, W., *Kontinuumsmechanik*, Stuttgart: Teubner, 1975.

[7] Belytschko, T., Liu, W.K. & Moran, B., *Nonlinear Finite Elements for Continua and Structures*, New York: John Wiley & Sons, 2000.

[8] Bouvard, D. & Stutz, P., Experimental study of rheological properties of a sand using a special triaxial apparatus. *Geotechnical Testing Journal* (1986), 9(1):10–18.

[9] Brekhovskikh, L.M. & Goncharov, V., *Mechanics of Continua and Wave Dynamics*, 2nd ed., Berlin: Springer-Verlag, 1994.

[10] Budhu, M., Nonuniformities imposed by simple shear apparatus. *Canadian Geotechnical Journal* (1984), 21(1):125–137.

[11] Callisto, L. & Calabresi, G., Mechanical behaviour of a natural soft clay. *Géotechnique* (1998), 48(4):495–513.

[12] Callisto, L. & Rampello, S., Shear strength and small-strain stiffness of a natural clay under general stress conditions. *Géotechnique* (2002), 52(8): 547–560.

[13] Casey, B., The consolidation and strength behavior of mechanically compressed fine-grained sediments, PhD thesis, Massachusetts Institute of Technology, 2014.

[14] Cnudde, V. & Boone, M.N., High-resolution X-ray computed tomography in geosciences: A review of the current technology and applications. *Earth-Science Reviews* (2013), 123:1–17.

[15] Davis, R.O. & Selvadurai, A.P.S., *Plasticity and Geomechanics*, Cambridge: Cambridge University Press, 2002.

[16] Desrues, J., Zweschper, B. & Vermeer, P.A., *Database for tests on Hostun RF sand*. Institutsbericht 13, Stuttgart: Universität Stuttgart, 2000.

[17] Desrues, J., Chambon, R., Mokni, M. & Mazerolle, F., Void ratio evolution inside shear bands in triaxial sand specimen studies by computed tomography. *Géotechnique* (1996), 46(3):529–546.

[18] Desrues, J., Lanier, J. & Stutz P., Localization of the deformation in tests on sand sample. *Engineering Fracture Mechanics* (1985), 21(4):909–921.

[19] Desrues, J., Andó, E., Bésuelle, P., et al., Localisation precursors in geomaterials? In E. Papamichos et al. (eds.), *Bifurcation and Degradation of Geomaterials with Engineering Applications*, Springer Series in Geomechanics and Geoengineering, Springer, 2017, pp. 3–10. doi: 10.1007/978-3-319-56397-8_1

[20] Desrues, J., Agrilage, A., Caillerie, D., et al., From discrete to continuum modelling of boundary value problems in geomechanics: An integrated FEM-DEM approach. *International Journal for Numerical and Analytical Methods in Geomechanics* (2019), 43(5):1–37.

[21] Doherty, J.P., Gourvenec, S. & Gaone, F.M., Insights from a shallow foundation load-settlement prediction exercise. *Computers and Geotechnics* (2018), 93:269–279.

[22] Falk, G., *Theoretische Physik, Band II Thermodynamik*, Berlin: Springer, 1968.

[23] Ferretti, E., On nonlocality and locality: Differential and discrete formulations. In *XVII National Conference Italian Group of Fracture*, Bologna, Italy, June 16–18, 2004.

[24] Feynman, R.P., *What Do You Care What Other People Think?*, New York: W. W. Norton 1988.

[25] Fellin, W., Abschätzung der Standsicherheit von annähernd unendlich langen Kriechhängen. *Geotechnik* (2011), 34(1):22–31.

[26] Fellin, W. & Ostermann A., The critical state behaviour of barodesy compared with the Matsuoka-Nakai failure criterion. *International Journal for Numerical and Analytical Methods in Geomechanics* (2013), 37(3):299–308. doi: 10.1002/nag.1111

[27] Gardner, W.R., Mathematics of isothermal water conduction in unsaturated soil. http://onlinepubs.trb.org/Onlinepubs/sr/sr40/sr40-009.pdf

[28] Glasstone, S., Laidler, K.J. & Eyring, H., *The Theory of Rate Processes*, New York: McGraw-Hill, 1941.

[29] Godunov, S.K., *Uravneniya Metematicheskoy Fiziki*, Moscow: Nauka, 1971.

[30] Goldscheider, M., Grenzbedingung und Fließregel von Sand. *Mechanics Research Communications* (1976), 3:463–468.

[31] Goldscheider, M., Der Erdruhedruckbeiwert K_0 von Reibungsböden – Materialgesetz und Bestimmung aus einem Standard-Triaxialversuch. *Geotechnik* (2020), 43:84–96. doi: 10.1002/gete.202000001

[32] Goldscheider, M., Mechanik des Kriechens von Böschungen und Hängen. *Geotechnik* (2014), 37(4):259–270.

[33] Gudehus, G., A comparison of some constitutive laws for soils under radially symmetric loading and unloading. In W. Wittke (ed.), *Proceedings of the Third International Conference on Numerical Methods in Geomechanics*, Rotterdam: Balkema, 4:1309–1324.

[34] Gudehus, G., Darve, F., & Vardoulakis, I., (eds.). *Results of the International Workshop on Constitutive Relation for Soils*, Rotterdam: Balkema, 1988.

[35] Hanisch, J., 'Wegweiser' auf dem Wege zu einem neuen Abschnitt in der Geschichte des Erd- und Grundbaus. *Bautechnik* (1995), 74(5):287–293.

[36] Hattab, M. & Hicher, P.-Y., Dilating behavior of overconsolidated clay. *Soils and Foundations* (2004), 44(4):27–40.

[37] Hight, D.W., McMillan, F., Powell, J.J.M., Jardine, R.J. & Allenou, C.P. Some characteristics of London clay. In T.S. Tan et al. (eds.), *Characterisation and Engineering Properties of Natural Soils*, Lisse: Swets & Zeitlinger, 2002, vol. 2, pp. 851–908.

[38] di Prisco, C. & Imposimato, S., Experimental analysis and theoretical interpretation of triaxial load controlled loose sand specimen collapses. *Mechanics of Cohesive-Frictional Materials* (1997), 2:93–120.

[39] Kitanidis, P.K., *Introduction to Geostatistics*, Cambridge: Cambridge University Press, 1999.

[40] Ishihara, K., Liquefaction and flow failure during earthquakes. *Géotechnique* (1993), 43(3):351–415.

[41] Khalili, N. & Khabbaz M.H., A unique relationship for χ for the determination of the shear strength of unsaturated soils, *Géotechnique* (1998), 48(5): 681–687.

[42] Knops, R.J. & Payne, L.E., *Uniqueness Theorems in Linear Elasticity*, Berlin: Springer, 1971.

[43] Kolymbas, D.A., generalized hypoelastic constitutive law. In *Proceedings of the Eleventh International Conference on Soil Mechanics and Foundation Engineering*, San Francisco, CA: A.A.Balkema, 1985, vol. 5, p. 2626

[44] Kolymbas, D. Computer-aided design of constitutive laws. *International Journal for Numerical and Analytical Methods in Geomechanics* (1991), 15(8):593–604.

[45] Kolymbas, D. & Bauer, E., Soft oedometer – A new testing device and its applications for the calibration of hypoplastic constitutive laws. *Geotechnical Testing Journal* (1993), 16(2):263–270.

[46] Kolymbas, D., Incompatible deformation in rock mechanics. *Acta Geotechnica* (2007), 2:33–40.

[47] Kolymbas, D., Barodesy: A new hypoplastic approach. *International Journal for Numerical and Analytical Methods in Geomechanics* (2012), 36:1220–1240. doi: 10.1002/nag.1051

[48] Kolymbas, D., Wagner P. & Blioumi A., Cavity expansion in cross-anisotropic rock, *International Journal for Numerical and Analytical Methods in Geomechanics* (2012), 36(2):128–139.

[49] Kolymbas, D., Barodesy: A new constitutive frame for soils. *Géotechnique Letters* (2012), 2:17–23.

[50] Kolymbas, D., Barodesy as a novel hypoplastic constitutive theory based on the asymptotic behaviour of sand. *Geotechnik* 35 (2012), 3:187–197.

[51] Kolymbas, D., Barodesy: The next generation of hypoplastic constitutive models for soils. In G. Hofstetter (ed.), *Computational Engineering*, Berlin: Springer International Publishing, 2014, pp. 43–56. doi: 10.1007/978-3-319-05933-4_2

[52] Kolymbas, D., Introduction to barodesy. *Géotechnique* (2015), 65(1):52–65.

[53] Kolymbas, D. & Medicus, G., Genealogy of hypoplasticity and barodesy. *International Journal for Numerical and Analytical Methods in Geomechanics* (2016), 40:2532–2550.

[54] Kolymbas, D., Barodesy with a new concept for critical void ratio. *Geotechnik* (2021), 3:166–177.

[55] Kolymbas, D., Bifurcation analysis for sand samples with a non-linear constitutive equation. *Ingenieur-Archiv* (1981), 50:131–140.

[56] Kolymbas, D. & Rombach, G., Shear band formation in generalized hypoelasticity, *Ingenieur-Archiv* (1989), 59:177–186.

[57] Kolymbas, D., *Geotechnik, Bodenmechanik, Grundbau und Tunnelbau*, 5th ed., Berlin: Springer, 2019.

[58] Kolymbas, D. & Bathaeian, I., Numerically obtained vortices in granular media. *International Journal for Numerical and Analytical Methods in Geomechanics* (2019), 43(16):2512–2524.

[59] Kozicki, J., Niedostatkiewicz, M., Tejchman, J. & Muhlhaus, H.B. Discrete modelling results of a direct shear test for granular materials versus FE results. *Granular Matter* (2013), 15(5):607–627.

[60] Krieg, S. & Goldscheider, M., Bodenviskosität und ihr Einfluss auf das Tragverhalten von Pfählen. *Bautechnik* (1998), 75:806–820.

[61] Krieg, S., Viskoses Bodenverhalten von Mudden, Seeton und Klei, Veröffentlichungen des Institutes für Bodenmechanik und Felsmechanik der Universität Fridericiana in Karlsruhe, 2000.

[62] Kuntsche, K., Materialverhalten von wassergesättigtem Ton bei ebenen und zylindrischen Verformungen, Veröffentlichungen des Institutes für Bodenmechanik und Felsmechanik der Universität Fridericiana in Karslruhe (1982), Heft 91.

[63] Kuznetsov, W.W. and Sher, E.N., The principle of homogeneous fracture of solids (in Russian), Doklady Akademii Nauk SSSR, 1976, 226(2):321–323.

[64] Landau, L.D. and Lifschitz, E.M., *Elastizitätstheorie*, Berlin: Akademie Verlag, 1966.

[65] Lanczos, C., *The Variational Principles of Mechanics*, New York: Dover Publications, 1970.

[66] Leinenkugel, H.J., Deformations- und Festigkeitsverhalten bindiger Erdstoffe. Experimentelle Ergebnisse und ihre physikalische Deutung. Veröffentlichungen des Instituts für Bodenmechanik und Felsmechanik der Universität Fridericiana in Karslsruhe, 1976.

[67] Lesne, A. & Laguës, M., *Scale Invariance*, New York: Springer, 2012.

[68] Love, A.E.H., *A Treatise on the Mathematical Theory of Elasticity*, Cambridge: Cambridge University Press, 1892.

[69] Nova, R. Plasticity theory treatment of geotechnical problems. In U. Smoltczyk (ed.), *Grundbau - Taschenbuch*, Berlin: Ernst und Sohn, 2001, 6th ed. part I, pp. 307–346.

[70] Nova, R., Controllability of the incremental repsonse of soil specimens subjected to arbitrary loading programmes. *Journal of the Mechanical Behavior of Materials* https://doi.org/10.1515/JMBM.1994.5.2.193

[71] Mandel, J., Consolidation des sols (Étude Mathematique). *Géotechnique* (1953), 3(7):287–299.

[72] Miura S. & Toki, S., A sample preparation method and its effect on static and cyclic deformation – strength properties of sand. *Soils and Foundations* (1982), 22(1):61–77.

[73] Morgenstern, N.R. & Tchalenko J.S., Microscopic structures in kaolin subjected to direct shear. *Géotechnique* (1967), 17(4):309–328.

[74] Muir Wood, D., *Soil Behaviour and Critical State Soil Mechanics*, New York: Cambridge University Press, 1990.

[75] Niemunis, A. & Herle, I., Hypoplastic model for cohesionless soils with elastic strain range. *Mechanics of Cohesive-Frictional Materials* (1997), 2:279–299.

[76] O'Donovan, J., O'Sullivan, C. & Marketos, G., Two-dimensional discrete element modelling of bender element tests on an idealised granular material. *Granular Matter* (2012), 14(6):733–747.

[77] Palmer, A. & Pearce, J., Plasticity theory without yield surfaces. In Palmer, A. (ed.), *Symposium on Plasticity and Soil Mechanics* (1973), Cambridge, England, pp. 188–200.

[78] Park, J. & Santamarina, J.C., Sand response to a large number of loading cycles under zero-lateral-strain conditions: evolution of void ratio and small-strain stiffness. *Géotechnique* (2019), 69(6):501–513. (https://doi.org/10.1680/jgeot.17.P.124)

[79] Peck, R.B., Advantages and limitations of the observational method in applied soil mechanics. *Géotechnique* (1969), 19(2):171–187.

[80] Pestana, J.M., Whittle, A.J. & Gens, A., Evaluation of a constitutive model for clays and sands: Part II – clay behaviour, *International Journal for Numerical and Analytical Methods in Geomechanics* (2002), 26:1123–1146. doi: 10.1002/nag.238

[81] Peters, J.F. & Walizer, L.E., Patterned nonaffine motion in granular media. *Journal of Engineering Mechanics* (2013), 139(10):1479–1490.

[82] Podio-Guidugli, P. & Favata, A., *Elasticity for Geotechnicians*, Berlin: Springer International Publishing, 2014.

[83] Polubarinova-Kotschina, P.Ya., *Teoriya Dvizheniya Gruntovykh Vod*, Moscow: Nauka 1977.

[84] Popov, L.V., *Kontaktmechanik und Reibung*, Berlin: Springer, 2009.

[85] Popper, K., Die offene Gesellschaft und ihre Feinde, Tübingen: Mohr, 7. Auflage 1972.

[86] Poulos, H.G. & Davis, E.H., *Elastic Solutions for Rock and Soil Mechanics*, New York: Wiley, 1974.

[87] Prandtl, L., *Führer durch die Strömungslehre*, Braunschweig: Vieweg, 1965.

[88] Prandtl, L., Ein Gedankenmodell zur kinetischen Theorie der festen Körper, *ZAMM* (1928), 8(2):85–106.

[89] Ruina, A., Slip instability and state variable friction laws. *Journal of Geophysical Research* (Dec. 10, 1983), 88:359–370.

[90] Saada, A. & Bianchini, G., (eds.), *International Workshop on Constitutive Equations for Granular Non-Cohesive Soils*, Rotterdam: Balkema, 1987.

[91] Schofield, A. & Wroth P., *Critical State Soil Mechanics*, London: McGraw-Hill, 1968.

[92] Schofield, A.N., Interlocking, and peak and design strengths, *Géotechnique* (2006), 56(5):357–358.

[93] Schubert, H., *Kapillarität in porösen Feststoffsystemen*, Berlin: Springer, 1982.

[94] Schweiger, H.F., Results from two geotechnical benchmark problems. In Cividini, A. (ed.), *Proc. 4th Europ. Conf. Num. Methods in Geotechnical Engineering*, Wien, New York: Springer 1998, 645–654.

[95] Skempton, A.W., Effective stress in soils, concrete and rock. *Proceed. Conf. on Pore Pressure and Suction in Soils*, London: Butterworth, 1960, 4–16.

[96] Sokolovski, V.V., *Statics of Granular Media*, New York: Pergamon Press, 1965.

[97] Sommerfeld, A., *Mechanik der deformierbaren Medien*, Leipzig: Akademische Verlagsgesellschaft, 1964.

[98] Strubecker, K., Einführung in die Höhere Mathematik, Band I, Munich: Oldenbourg, 1956.

[99] Strubecker, K., Einführung in die Höhere Mathematik, Band II, Munich: Oldenbourg, 1967.

[100] Sultan, N., Cui, Y.-J. & Delage, P., Yielding and plastic behaviour of Boom clay. *Géotechnique* (2010), 9:657–666. doi: 10.1680/geot.7.00142

[101] Tamagnini, C., Viggiani G. & Chambon, R., Some remarks on shear band analysis in hypoplasticity. In H.-B. Mühlhaus et al. (eds.), *Bifurcation and Localisation Theory in Geomechanics*, Lisse: Swets & Zeitlinger (2001), pp. 1–9.

[102] Tatsuoka, F., di Benedetto, H., Enomoto, T., et al., Various viscosity types of geomaterials in shear and their mathematical expression. *Soils and Foundations* (2008), 48(1):41–60.

[103] Taylor, D.W., *Fundamentals of Soil Mechanics*, New York: Wiley, 1966.

[104] Terzaghi, K. von, Relation between soil mechanics and foundation engineering. In *Proceed. Intern. Conf. SMFE*, Vol.III, Cambridge, MA: Harvard University Press, 1936, 13–18.

[105] Thornton, C. & Zhang, L., A numerical examination of shear banding and simple shear non-coaxial flow rules. *Philos. Mag.* (2006), 86(21–22): 3425–3452.

[106] Tonti, E., *On The Formal Structure of Physical Theories*, Milan: Istituto di Matematica del Politecnico di Milano 2004.

[107] Topolnicki, M., *Observed Stress-Strain Behaviour of Remoulded Saturated Clay and Examination of Two Constitutive Models*. Veröffentlichungen des Institutes für Bodenmechanik und Felsmechanik der Universität Fridericiana in Karlsruhe (1987), Heft 107.

[108] Tordesillas, A., Pucilowski, S., Walker, D.M., Peters, J.F. & Walizer, L.E., Micromechanics of vortices in granular media: connection to shear bands and implications for continuum modelling of failure in geomaterials. *International Journal for Numerical and Analytical Methods in Geomechanics* (2014), 38:1247–1275. doi: 10.1002/nag.2258

[109] Truesdell, C.A. & Noll, W., The non-linear field theories of mechanics. In S.Flügge (ed.), *Encyclopedia of Physics*, Vol. IIIc. Berlin: Springer, 1965.

[110] Truesdell, C. & Toupin R.A., Principles of classical mechanics and field theory, In S. Flügge (ed.), *Encyclopedia of Physics*, Vol. III/1. Berlin: Springer, 1960.

[111] Truesdell, C., *An Idiots Fugitive Essays on Science*, New York: Springer, 1984.

[112] Timoshenko, S.P. & Goodier, J.N., *Theory of Elasticity*, 3rd ed., New York: McGraw-Hill, 1970.

[113] Verdugo, R. & Ishihara, K., The steady state of sandy soils. *Soils and Foundations* (1996), 36(2):81–91.

[114] Verruijt, A., *Soil Mechanics*, Delft University of Technology, 2012, geo.verruijt.net/software/SoilMechBook2012.pdf

[115] Wichtmann T., Soil behaviour under cyclic loading - experimental observations, constitutive description and applications, Veröffentlichungen des Instituts für Bodenmechanik und Felsmechanik am Karlsruher Institut für Technologie (KIT) (2016), Heft 181. www.torsten-wichtmann.de

[116] Wilmański, K., *Thermomechanics of Continua*, Berlin: Springer, 1998.

[117] Witham, G.B., *Linear and Nonlinear Waves*, New York: John Wiley & Sons, 1973.

[118] Wolf, J.P. & Deeks, A.J., *Foundation Vibration Analysis: A Strength of Materials Approach*, New York: Elsevier, 2004.

[119] Wolffersdorff, P.A. von, Feldversuch an einer Spundwand in Sandboden: Versuchsergebnisse und Prognosen. *Geotechnik* (1994), 17(2):73–83.

[120] Yong, R. & Darve, F., (eds.), *Workshop on Limit Equilibrium, Plasticity and Generalized Stress-Strain in Geotechnical Engineering*, McGill University, Montreal: ASCE, 1980.

Index